THE VITAL LINK BETWEEN WATER AND HEALTH

PURE WATER FOR BETTER LIVING

ELDON C. MUEHLING

Pure Water for Better Living

Copyright © 2015 by Pure and Secure, LLC

ISBN: 978-0-9962043-0-9

PRAISE FOR AUTHOR ELDON MUEHLING

"A real man with integrity. A man glowing with love. Also my favorite teacher of all times. His knowledge and down to earth style of teaching really moved me to become a better educator of distilled water."

— *Denny Mahoney, Midwest Health and Wellness Association*

"Eldon is my 'go to guy' for tough science questions about water. It's clear and in everyday layman's terms. It's just science made simple. Thanks Dr. Water!"

— *Reggie Nisbett, President, Greater Texas Water Company*

"When I first started with Pure Water in 1975, Eldon was the one who answered all my questions and helped me with every question I have asked since. We have become good friends through the years and I respect him and his scientific knowledge."

— *Gary McKenzie, Independent Distributor*

"It was the year 2001 when we contacted Pure Water in Lincoln, Nebraska. We had an idea of installing a central pure water system in a high-rise building with 225 residential apartments. Needless to say this would be an extremely costly endeavor. After speaking to Al Meder at Pure Water we were convinced that they would have the equipment necessary to facilitate our project. We had a real estate developer willing to be our partner in this project, the only thing missing at that point was the courage to continue not knowing if this could be done on such a large scale. Thankfully Eldon was an on staff consultant at the time for Pure Water and had so much knowledge of pure water systems that after speaking with him on a daily basis for quite some time, he helped us muster up the courage to proceed with the project. Eldon was always happy to take our calls and totally convinced us that distilled water would be the best application for this endeavor. I must say that he was partially responsible for our success. Large-scale central pure water systems in apartment buildings are now catching on in the Northeast. Thank you Eldon."

— *Phil Festa and Laurrie Cozza, H2onlywater.com*

"It's a total delight learning chemistry from Eldon. He makes teaching science an art form."

— *Kevin T Smith, Worry-Free Water*

"I first met Eldon when I experienced kidney stones and found a business card from one of Eldon's salesmen. That led me to Eldon Muehling and the wonderful company he represents—Pure and Secure. That was in 1979 and Eldon, with his degree and post-degree schooling in chemistry, was a real messenger of truth in water. Now, some 35 years later, I can't even begin to evaluate the benefit of his water knowledge to three generations of my family with his cut-to-the-truth answers to every query about water we've had. With Eldon's help, I personally have managed to lead many, many people to the best drinking water there is: Distilled!"

— *Mark Wickering, Independent Distributor*

"Eldon is the best read authority I know on water with all the facets and chemistry surrounding H2O. He is a 'horn of plenty' on the subject. He has been wonderful to have as a reference and a friend."

— *Norm Grable, Independent Distributor*

DEDICATION AND PREFACE

This edition of *Pure Water for Better Living* (a revised version of the previously published *Pure Water Now)* is dedicated to Al Meder, President and owner of Pure & Secure, LLC in celebration of the company's 47th anniversary. Without the vision of Al Meder, the Pure Water™ Brand of Water Distiller may not exist today.

It has been 20 years since I wrote the original version of this book. I am grateful for the many wonderful Pure Water experiences that I have had and shared with our dealers and customers over this time. Much has changed in the world and in the water business. It is because of this that I am revising the original version of this book to update readers on many of these changes.

I would also like to take this opportunity to extend my heart-felt thanks to Courtney (Meder) Lawyer for her many hours of editing and putting the finishing touches on this revised and updated edition of *Pure Water for Better Living*.

One thing for sure is that the water today is no better than it was 20 years ago. In fact, by most measures it is now even worse. The water problems of yesterday have grown in complexity and extend to more and more of us each day. Man-made chemicals, which are so much a part of our everyday life now, end up in our drinking water. Pharmaceuticals and personal care products are prime examples of some of the new contaminants that have invaded our water supply. Also, the concern for radioactive contamination is considerably greater today than it was 20 years ago, in large part due to the melt down of the nuclear power plant in Fukushima, Japan after a large earthquake in 2011. To some extent, we all still take water for granted… that is until we don't have any or the quality of it is so bad we can't stand to pass it by our lips. Fortunately, for those of us who own a water distiller, even the worst water can be converted to safe and beneficial pure water.

I encourage you to take the need for pure drinking water personally and pursue the appropriate remedy by either owning or renting a pure water distiller. Your body will thank you in many ways. For more information about the health benefits of pure water, please visit these websites:

www.MyPureWater.com

www.MyAquaNui.com

Eldon C. Muehling

Lincoln, Nebraska

FOREWORD

Eldon Muehling has been a personal friend of mine for 30 years. Eldon is a unique combination of scientist, teacher and distilled water guru. Over several decades he has been known and loved as "Dr. Water". In that time he has personally offered advice to thousands of people who have sought his expertise on a whole range of topics relating to water.

Eldon is a strong believer that one's health is tied to the quality of water. This is why he has been such a strong advocate of drinking distilled water and has convinced so many families to follow his advice.

One may wonder how much could change with water. It's been more than 20 years since the publication of his first edition of *Pure Water Now* and in that time there has been so much more publicity on the dangers of certain contaminants in drinking water. There have also been some significant examples of water pollution on a massive scale—the 1993 *Cryptosporidium* contamination in Milwaukee, Wisconsin, where more than 400,000 drank contaminated water and thousands were hospitalized. Many today are surprised to learn that more than 100 people died in that incident. Then there are the much more recent examples of the contamination of the Elk River in West Virginia and the blue-green algae occurrences in Lake Erie. Both left hundreds of thousands without drinking water for several days.

The problem is that water can look and smell clean and pure while in actuality it may not be. Eldon, in this latest edition of his writing, covers many aspects of the new issues with water.

Even if you have read Eldon's earliest book, I know you will be blown away by this one. Water is so critically important in our lives, yet I would guess that the vast majority of people are unknowingly ignorant on this topic.

If you are not already part of the slim minority that understands the importance of water, I'm pleased to say that after reading this book, you will join this group of people who understand water in depth.

Al Meder

President

Pure & Secure, LLC

Lincoln, Nebraska

TABLE OF CONTENTS

Introduction:
The Unseen Threat

n 1993 the Howard Avenue Water Purification Plant in Milwaukee, Wisconsin was contaminated by a protozoan cyst called *Cryptosporidium*. Newspapers at the time documented that more than 400,000 people of all ages became violently ill, several thousand were hospitalized and more than 100 died in one of the largest outbreaks of water-borne disease in the United States in recent times.

Compare that to the recent fear and panic surrounding Ebola. From August 2014 through February 2015, according to the Centers for Disease Control, only four confirmed cases and two deaths actually occurred on US soil from the virus.

Just look at the numbers: The Milwaukee water contamination directly impacted 100,000 times as many people and caused 50 times the number of deaths as Ebola. Yet, I suspect if you asked people if water contamination or Ebola posed a greater potential threat to their health, most would say Ebola.

The truth is water can be a major health risk. First, most of us don't drink enough of it. By the time you're feeling thirsty, you have already waited a little too long to replace some of the vital water your body loses through daily exertion. One telling study of athletes revealed that the amount of water required for peak performance was equal to the amount of water they desired, plus one-third. We must actually drink more pure water than our thirst demands if we wish to get the maximum performance from our bodies.

Secondly, the water we do drink can contain chemicals, viruses, bacteria and inorganic substances that pose a threat to our health. We can't always taste, see or smell the toxins that may be lurking in our drinking water. Even some of the 47 chemicals often used to treat water supplies in the US can be harmful to humans. As use of industrial, agricultural and even household chemicals skyrockets, municipal water supplies are more at risk. Water contaminated with radioactive waste from the 2011 Fukushima nuclear disaster has even reached our shores. Most of us don't know what's in the water we drink, cook with and even use for mixing baby formula.

This news doesn't make the headlines and capture public attention the way the Ebola virus did, but perhaps it should. We tend to take our drinking water for granted. We don't stop to consider the vital role water plays in the health of the human body and we underestimate the risks of water-borne contaminants. This book will help you understand the science behind water, update you on the safety of water supplies, and give you the basic information you need to make wise decisions about the water your family consumes.

Chapter 1:

Water is Life

W e all consider the human body to be unique and it certainly is. In simple terms, it is actually a highly specialized, porous blob of water. This watery bag is formed into shape by a protective skin with a little hair (more or less) growing from it. The amount of water in the human body averages about 65% by weight and about 75% by volume. It varies considerably from person to person and even from one part of the body to another. A lean healthy man may be as much as 70% water by weight. On the other hand, the average woman, because of her larger proportion of water-poor fatty tissues, may be as little as 52% water. The percentage of water in a woman also oscillates with her menstrual cycle. A newborn baby is about 90% water, unless of course it is a chubby baby, in which case it is less.

Water is not only the most abundant compound in the human body but it is also the most abundant component in the human diet. We talk of breathing air, but what all living things really do is to breathe oxygen dissolved in water. Both the quantity and the quality of the water you ingest profoundly determine the quality of your tissues and their performance and resistance to disease. The purer the water, the better!

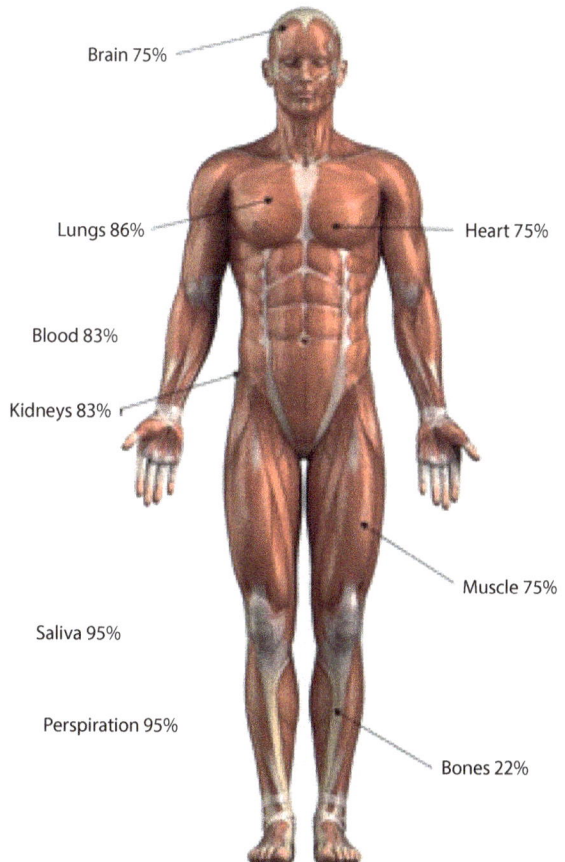

Brain 75%

Lungs 86%

Heart 75%

Blood 83%

Kidneys 83%

Muscle 75%

Saliva 95%

Perspiration 95%

Bones 22%

HUMAN CELLS ARE 60% WATER

Every cell, the basic unit of the human body and every other living organism, contains about 60% by weight of this vital substance. No two cells in the body are actually touching as there are two large classes of body water: extra-cellular water, which is water that is outside of the cells; and intra-cellular water, which is water that is inside the cells. A precise balance of water outside and inside cells is essential for the maintenance of life. This is magnificently accomplished with the proper ratio of sodium and potassium intake. Trans-cellular water is water participating in the transport activity of specialized cells such as salivary and liver cells, spinal fluid, and fluids of the eye.

Tests of athletes show failure to replenish your body's water will result in your performance suffering immediately. If a muscle is dehydrated by as little as 3%, there is a 10% loss of contractile strength and as much as an 8% loss of speed. In a test conducted at the Colgan Institute in Encinitas, California a number of years ago, a group of athletes were made to run a 5,000 meter race after fasting overnight. Their times were off an average of almost one minute in a run that ordinarily took less than 20 minutes. Failure to maintain their water level for only a few hours had cost these athletes a 5% loss of performance. That could be the difference between winning the race and last place.[1]

ALL ANIMALS AND PLANTS ARE LARGELY WATER

Not only is an adequate water supply essential for human beings, but it is also a matter of life and death for every form of animal and plant life, from the lowliest amoeba to the tallest redwood. Water is more than

---

1   Dr. Michael Colgan, *Optimum Sports Nutrition*. Ronkonkoma, New York: Advanced Research Press, 1993.

the first requisite of our lives—it is a most thrilling subject. Its influence touches us constantly, every moment of every hour, from the first spark of life to the final breath and beyond.

Water is found in nearly everything including stones and minerals. Even coal has a small percentage. Water constitutes about 80% of all living matter. Every living animal organism depends on water for at least 50% of its body weight, and must constantly replenish the water lost through evaporation and excretion. Fortunately, each has evolved an effective means of satisfying this need.

Most of the foods we depend on to sustain us are largely water: whole milk being 87% water, eggs 65%, beef 62%, even bacon is 22% water. Consider the following table which was compiled from a number of scattered but reliable sources:

PERCENTAGE OF WATER IN COMMON ANIMALS		PERCENTAGE OF WATER IN COMMON FOOD PLANTS	
Weevil (insect)	48%	Sunflower seed	5%
Rat	65%	Apple Seed	10%
Fish	67%	Corn Kernel	70%
Elephant	70%	Apple (ripe)	80%
Chicken	74%	Pineapple (ripe)	87%
Frog	78%	Carrot	90%
Earthworm	80%	Tomato (ripe)	90%
Jellyfish	95%	Watermelon	97%

Fruits and vegetables tend to increase their water content as they ripen. That is why they taste so much better when they are vine-ripened or picked and used fresh from the garden. Fruits and vegetables do not die when they are picked. They do, however, begin to lose water and quality as soon as they are harvested. It is very important they be stored properly and used as soon as possible.

To me, it's an absolute miracle that an apple seed, which is only 10% water, will eventually produce a nice red delicious apple, which is 80% water! Pure water is life.

Animals, unlike plants, must maintain a fairly rigid percentage of water in their bodies in order to survive. Most animals have adapted their physical needs to match the water levels of their environment. The desert-dwelling kangaroo rat, for example, can get along on an absolute minimum of water, but the jellyfish must remain immersed in water at all times. A donkey can survive in a hot desert for four days but would lose up to 30% of its body weight. The man riding the donkey will be dead when his water loss reaches 15%. How much would a glass of cool, pure water (any water for that matter) be worth to that man?

WHERE ARE ALL OF THE ANIMALS?

Fifty years ago there were considerably more bees and other insects, frogs, birds, fish… the list goes on. Where have they gone? What has changed? Could pollution have caused their disappearance? It is estimated that human beings have found or manufactured *more than 50 million chemicals* and the speed with which we develop new ones is also increasing. According to the American Chemical Society, it took 33 years to get the first ten million chemicals registered, but only nine months to get the last ten million chemicals registered! Wow! Just think about that for a moment.

We all use many chemicals in our everyday lives. We use them for our personal use and medical care. We use them when building our homes and later to repair them. We use them in our vehicles and on our streets, yards and fields. We use them to kill weeds as well as insects and other pests. Add to that the chemicals used by businesses, manufacturers and those who produce our food. Many of these chemicals will eventually end up in the air we breathe and the water we drink. For some of them, it will be sooner than we think.

Perhaps even more disturbing is the fact that the chemical concentrations are also rising, and fairly rapidly in many cases, especially in those areas where the populations are increasing. The concentration of chemical pollutants in the air, VOCs as they are called (volatile organic compounds), are also on the rise. Again, common sense points to humans as the primary cause. It will take common sense and practical solutions to reverse the damage already done to our environment. There are many things that we can do but first and foremost, we should develop new chemicals that are safer and use fewer of them wherever and whenever possible.

WATER DOMINATES HISTORY

Water has dominated mankind's evolution and history, as well as that of all other living things. Today, water, more than any other substance, is dominating man's future progress as well. While we often tend to take water for granted, it is obviously a critical issue in times of floods, drought, widespread contamination of community water supplies, or war. Even future community growth is often determined by the accessibility of plentiful and safe water supplies.

Looking back over the course of history, one will quickly discover that more disputes have been started and wars fought over water than over any other commodity. Not land, not oil, not even gold and other riches. Why? Because water is absolutely essential to life, and most of us recognize it. Land, gold and oil aren't, even though some short-sighted people may foolishly think they are. One can easily predict this interrelationship will only escalate in the future. Yet, even though water is as essential to us as the air we breathe, it is just as vaguely understood, and sorry to say, even abused. Water affects our health, our prosperity, our joy of living. We sure as heck ought to be interested in this. It behooves us to learn a lot more about this wonderful and fascinating substance called water. Our very existence depends on it!

Chapter 2:

Understanding Water

"**W**ater," says the humorist "is something they put under bridges." That is about as seriously as some of us regard it— except the fellow who has just upset the boat! Most of us know water in small quantities to be a colorless liquid that descends from clouds as rain and forms oceans, rivers, streams, lakes and ponds. In large bodies of water, water appears to be blue; however pure water is a transparent, colorless, tasteless, and odorless compound.

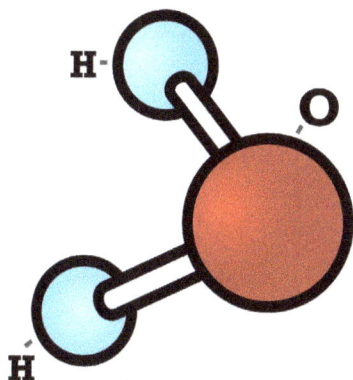

Water is composed of two invisible gases: hydrogen (11.11% by weight) and oxygen (88.89% by weight), but it displays the properties of neither. Hydrogen, for example, is a highly flammable and explosive gas, while oxygen is absolutely necessary for combustion. Yet these two gases, combined in the form of water, readily extinguish fire! How does it do that? By absorbing a large amount of heat from the fire and surrounding the burning structure so as to exclude the oxygen of the air, without which the fire cannot continue. We'll get into a little more of the chemistry of water later. For now, we're going into a brief discussion of water's physical nature.

PURE WATER SET THE STANDARDS

The freezing points and boiling points of pure, distilled water were used to establish both the Centigrade (Celsius) and Fahrenheit temperature scales. At sea level or standard pressure, pure water freezes at 0° Centigrade (Celsius) and boils at 100° Centigrade (Celsius). The corresponding fixed points on a Fahrenheit temperature scale are 32° F and 212° F.

Furthermore, the specific heat of pure water is used to define both the calorie and a B.T.U. (British thermal unit). A calorie is defined as the amount of heat required to raise the temperature of one gram of pure water one degree centigrade. Correspondingly, a B.T.U. is defined as the amount of heat required to raise the temperature of one pound of pure water one degree Fahrenheit.

Pure water is at its maximum density (molecules closest together) at 4° Centigrade (39.2° F). It is at this temperature that pure water serves as the standard for the specific gravity of solids and other liquids. At this same temperature, one cubic centimeter (milliliter) of pure water is used to define one gram, the metric unit of mass.

NOTE: The significance of the previous paragraph is that water expands (about 9%) as it freezes and is therefore lighter than the water in which it is freezing. Consequently, the ice floats to the top. That's why ninety-percent (90%) of an ice cube or iceberg is below the surface of the water. This also allows aquatic life to survive in large bodies of frozen water, since the temperature at the bottom will be a nice toasty 39.2° F, even when the outside air temperature may be well below freezing! Likewise, the bottom of the pond is a nice cool place for a bass to be in the middle of July. Any good fisherman knows this and casts his lines accordingly.

This image depicts water in all three states: solid liquid and gas.

WATER IS A MAVERICK

Consider if water behaved like every other liquid. Ice would have settled to the bottom as it froze and our open bodies of water would have frozen solid ages ago! Life as we know it would most certainly not exist. We can be thankful that water is such a maverick. When astronomers search for life in the universe, they are in fact actually looking for water.

Water is the only substance on earth that is naturally present in the solid, liquid and gaseous states within the usual range of temperatures on the planet (see image). Even more strangely, water can even exist in all three forms at the same temperature! Again, unique!

ALL HUMAN FUNCTIONS DEPEND ON WATER

There is nothing even remotely similar to water. One example of this is the number of functions it carries out in the human body: breathing, digestion, growth, movement, elimination of waste products, heat dissipation, secretion, and all glandular activities. These can only be performed in the presence of watery solutions. In fact, water affects every facet of our physiology.

Water is the most efficient cooling agent known because it takes so much heat just to vaporize one gram. It requires approximately 540 calories, even when the water is at its boiling point. Yet, it evaporates easily at body temperatures. We are reminded of this property of water whenever we leave the swimming pool dripping wet. Teeth chattering, we quickly grab for a dry towel in order to avoid getting too cold.

At other times, our bodies automatically release perspiration to maintain normal body temperature. This is why sweating, or should I say perspiring, on a hot day is a very good thing from a health standpoint. On a much larger scale, it is this cooling property of water, along with other physical factors, that influences the formation of ocean currents, the movement of winds and other aspects of weather and climate.

PURE WATER: THE UNIVERSAL SOLVENT

Liquid pure water is the best known solvent in existence. In fact, it is commonly referred to as the universal solvent. This means it can dissolve a greater number of substances than any other liquid. Water's staggering capacity to dissolve is illustrated by the fact that one gallon of pure water, which weighs a little over 8 pounds, can dissolve 70 pounds of ammonium nitrate, the fertilizer we commonly spread on our lawns, gardens and fields.

The great diversity of substances dissolved by water is illustrated by the fact that ocean water contains about half of the known elements.[2] Pure water is truly a maverick among other liquids in many ways. Water molecules, in contact with foreign substances, can be compared to cowboys cutting cattle from a herd…they force their way between clusters of particles, break them apart and hold them at bay.

---

2   H. U. Sverdrup, Martin W. Johnson and Richard H. Fleming, The Oceans, Their Physics, Chemistry, and General Biology. New York: Prentice-Hall, c1942 1942. http://ark.cdlib.org/ark:/13030/kt167nb66r/

ALL LIFE DEPENDS ON WATER

Without water's tremendous ability to dissolve other substances, nutrition of any type could not go on. For example, living micro-organisms depend on water to dissolve the substances they feed on. Likewise, unless the nutrients in the soil are in solution, the roots of the plants cannot absorb them. Human food must be dissolved in gastric juice, which is largely water, before it can enter the bloodstream for distribution to the individual cells. Absence of water from the diet will cause a quicker death than if every other dietary need is withheld. In short, there's nothing like water!

Chapter 3:

Thirst

Thirst serves as a very important mechanism for regulating the water intake of the body. It is not known whether the decreased flow of saliva is a result of reduced water in the blood, or in certain body cells, such as those in an area of the brain called the hypothalamus.

The hypothalamus is somehow sensitive to the concentrations of water and dissolved substances in the blood that bathes it. If the concentration of water is low, the sensation of thirst is experienced. One should recognize feelings of thirst as the cells, tissues and organs of your body crying out for more water. By the time you're feeling thirsty, you have already waited a little too long to replace some of the water your body has lost. Water, other beverages, and foods that contain large amounts of water are consumed to relieve this sensation. Pure water does the best job of quenching one's thirst

At any rate, nature's built-in thirst mechanism is a wonderful protective device, and we find we get thirsty for water much more often than we get hungry for food. As a result, most of us swallow two to three quarts of water every day, half in food and half as free water.

Vegetarians will probably find their amount of free water to be somewhat less. Remember, we must replace the water we eliminate each day.

DECREASE YOUR WATER—INCREASE YOUR FAT AND VICE VERSA

Pure water is the single most important ingredient to good health and a proper diet. Various studies have shown that a decrease in water intake will cause body fat to increase, while an increase in quality water can actually reduce body fat as it encourages the body's metabolism.

According to a study conducted by Brenda Davy, PhD, an associate professor of nutrition at Virginia Tech, drinking just two eight-ounce glasses of water before meals helps people lose weight. This study was presented at the 2010 National Meeting of the American Chemical Society in Boston. Davy said that results of the first randomized controlled intervention trial demonstrated that increased water consumption is an effective weight-loss strategy. She added that in earlier studies, middle-aged and older people who drank two cups of water right before eating a meal, ate 75 to 90 fewer calories during their meals.[3]

When the body gets less water, it perceives this as a threat to its survival and begins to hold on to every drop. This water is stored in the extra-cellular spaces (spaces between the cells) which leads to swelling. Water retention is overcome by giving your body what it needs: plenty of pure water.

WHY IS THE POTASSIUM/SODIUM RATIO SO IMPORTANT?

Both potassium and sodium are essential nutrients that your body needs to control fluid and electrolyte balance. Consume too much sodium and the body will tend to retain water. Consume too much potassium, on the other hand, and the body tends to become dehydrated. Studies have shown that the ratio between potassium and sodium is very important. Unfortunately most folks, Americans in particular, consume too much salt (sodium chloride) and therefore have a ratio that is too low. This can lead to hypertension (high blood pressure), stroke, kidney stones, osteoporosis, gastrointestinal tract cancers, asthma, exercise-induced asthma, insomnia, air sickness, high-altitude sickness, and Meniere's Syndrome (ear ringing). Researchers advise that the optimal potassium to sodium ratio should be greater than two to one (2/1). Eating more bananas, a good source of potassium, is only part of the

---

3   Dennis, E.A., Dengo, A.L., Comber, D.L., Flack, K.D., Savla, J., Davey, K.P., and Davy, B.M., "Water Consumption Increases Weight Loss During a Hypocaloric Diet Intervention in Middle-aged and Older Adults." Obesity (Silver Spring), 2010 Feb;18 (2):300-7. doi: 10.1038/oby.2009.235. Epub 2009.

answer. Avoiding drinking water with high levels of sodium is another part of the solution.

EARLY EXPERIMENTS WITH THIRST

In the early 1940s, Dr. G.C. Pitts conducted some very fascinating experiments with athletes at the Harvard University School of Public Health. In three separate experiments he showed what happened to these athletes who were exercising in a hot environment. The first experiment allowed the athletes to drink no water. During the second experiment, the athletes were allowed to drink as much water as they desired. In the third experiment, the amount of water they lost as sweat was measured and they were forced to drink this amount of water. The athletes were instructed to walk as long as they could on a treadmill at three and one-half miles per hour. At regular intervals, they were allowed five-minute rest periods during which their body temperatures were taken and recorded.

FIRST EXPERIMENT: During the first experiment (no water), their temperatures rose from 98.6° F to 102° F (the zone of impending exhaustion) in only three and one-half hours. When body temperatures rise above 102° F, physiological functions are impaired and collapse is inevitable.

SECOND EXPERIMENT: On a subsequent day, after the athletes had recuperated, they did the same experiment again except during rest periods they were allowed to drink as much water as they desired. Their body temperatures rose more slowly. It took six hours before they entered the zone of exhaustion (almost twice as long as before).

THIRD EXPERIMENT: Finally, again after the athletes had recuperated, they did the experiment a third time. This time the athletes were drinking as much water as they were losing in sweat. Amazingly, their body temperatures never even reached 101° F. After seven hours, when the experiment was ended, the athletes felt they could go on indefinitely!

Dr. Pitts discovered in the second experiment, when the athletes drank all the water they wanted (or thirsted for), they were actually drinking one-third less water than they lost in sweat. The amount of water they really needed was equal to the amount of water they desired, plus one-third. This and other experiments have shown that thirst is an unreliable indicator of the amount of water one should drink. We must force ourselves to drink more pure water than our thirst demands if we wish to get the maximum performance from our bodies.[4]

> NOTE: The next couple sections will get a bit on the mathematical side. They help to explain the importance of drinking adequate amounts of water for proper elimination. If such material is a bit much for you, skip the sections with an asterisk. Otherwise, read on.

WATER GAINS MUST EQUAL WATER LOST*

An average human being doing light work in a temperate climate loses nearly five pints of water each day. Therefore, he must replace five pints of water each day. Assuming he can get about half of his water from the food he is eating that day, he can take in two-thirds of a pint each meal. That's two pints total for the day from the food he eats, assuming he eats all three meals. Now here's something you may not know. When the human body metabolizes food, both energy and water are produced. Feces are the by-product.

Continuing with our example, this average man creates about three-fourths of a pint of water per day within his own body in the oxidizing

---

4   Philip Collins, "The Health Benefits of Drinking Water." Mother Earth News. July/August 1986.

of his food, notably the combustion of fat. Combined with the two pints of water from his food, this leaves two and one-fourth pints or so of water to be replaced from water or watery liquids-preferably pure distilled water. (2 pints + ¾ pint + 2 and ¼ pints = 5 pints.) Great! That works out just right.

TAKING A CLOSER LOOK AT WATER LOSS*

The daily five-pint water loss of this average man is made up of about one-fourth pint in his feces, nearly one pint from his lungs (the moisture in the air that he exhales), some two and one-half pints of urine, and one-half pints of urine, and one and one-fourth pints of perspiration. (¼ pint +1 pint + 2½ pints + 1¼ pints = 5 pints.) In unusual circumstances, as in the presence of diarrhea, vomiting, hemorrhage or excessive perspiration, there is a sudden excessive loss of water accompanied by a corresponding sudden and urgent thirst.

> NOTE: Some people think sweating occurs only during strenuous exercise. This is not so. Instead, it is customarily only visible during such exercise, when evaporation fails to keep pace with perspiration. Even a man at rest is losing more than one-half ounce of water through his skin every hour. That works out to be 12 ounces a day (three-fourths of a pint) even if you spent the last 24 hours sleeping in your bed! You better wake up now and then long enough to drink a glass of water.

WATER INPUT & OUTPUT ARE HIGHLY VARIABLE

The quantity of water excreted by the kidneys is almost in direct proportion to the amount of water taken into the body. The quantity eliminated by the kidneys varies daily. In certain physical irregularities, notably diabetes, as much as three gallons of water is eliminated in 24 hours. It is much better to have too much water in the body than not enough. It is the function of the kidneys to determine when the body's blood system has an excess of specific essential substances and furthermore to remove such excesses from the blood in an effort to maintain balanced blood chemistry. When our blood becomes too concentrated, our pituitary gland secretes a chemical substance called vasopressin. The blood carries this chemical to the kidneys where it brings about the return of as much water as possible to the blood.

Another function of the kidneys of paramount importance to the body's blood is the regulation of the body's acidic/alkaline balance. The balance between acidity and alkalinity in the body's fluids is governed largely by the relative number of hydrogen ions present. It is the kidney's responsibility to maintain the proper acid-base balance in the body by eliminating the appropriate amount of hydrogen ions by way of the urine. Certain cells lining the kidney tubules have the ability to eliminate ammonia, which also directly influences the body's acid-base balance.

THE ROLE OF THE KIDNEYS

The substances that the kidneys must regulate, besides water, include: glucose, carbonates, sodium, chlorides, potassium, sulfate, calcium, phosphate, and urea. With the exception of urea, the substances passing through the kidneys are crucial to normal body functions. Urea is a waste product destined for elimination in the urine. The kidneys will eliminate any surplus, with surprising speed. However, nothing but harm to the efficiency of the body will accrue when there is not enough water. Do your kidneys a big favor and drink more water. Pure water, that is.

The way our kidneys accomplish their functions is by filtering all of our blood through a highly precise set of duct systems known as nephrons. Our kidneys filter approximately 180 quarts of blood in a 24 hour period. Each molecule in the blood passes through a funnel checkpoint (nephron) where it is either directed back into the body's blood supply or directed to exit the body.

These many varied and vital functions of the kidneys proceed automatically and efficiently, as long as we give the body adequate supplies of water each day. How fortunate we are that we do not have to give conscious attention to the amount of water we should eliminate from our body each day. Furthermore, we do not have to calculate and count the number of sodium or potassium ions, calcium ions or phosphate radicals to ensure a sufficient number remain in our body. How wondrously we have been created! In examining the many wonderful operations our kidneys carry out, you no doubt will simply have to agree it makes awfully good sense to indulge your kidneys with as much water as they need, and the purer the water, the better. Your kidneys deserve it and so do your taste buds!

ON AVERAGE, YOU CAN LIVE ONLY TWO OR THREE DAYS WITHOUT WATER!

While the average human being can survive up to two months without solid food, it can survive only two or three days without water. Our bodies store food in the form of fat, but there is virtually no provision for the storage of water. In cases of serious and prolonged water deprivation, death is imminent. Not only do the kidneys fail and all kidney functions shut down, but the skin shrinks (much like a tomato baking in the sun), and begins to crack. Soon the mouth and tongue dry out. Next hallucinations set in, and ultimately with the loss of 10%-20% of the body's water, an agonizingly painful death will follow shortly. Literally, we must live in a running stream of water.

WE ALL NEED MORE PURE WATER NOW

This section may seem a bit crude to some of you but extremely important. Adequate water in the system is also necessary for comfortable bowel movements. Believe it or not, 45% to 55% of the mass of material you pass in your stool is microbial (bacteria and viruses!). When there is too little water in the system, it is used first for other, more vital functions, such as heat dissipation or balancing blood chemistry. The water that remains may not be adequate to materially aid elimination through the rectum. The stool will be hard and dry. Constipation, which many authorities call the disease of diseases, is the unfavorable result. Improper elimination of bowel waste can be one of the results of some eating disorders and can cause a number of other health problems. Of course, diarrhea will step up water loss greatly and is generally avoided by adequate intake of plant fiber. Many of the residents in nursing homes are constipated as a result of not drinking enough water.

TYPE OF WORK AFFECTS WATER EXCHANGE

Your body is like a water-cooled engine. Neither you nor an engine can perform very long or very well on a hot day without enough water. The type and conditions of an occupation will have a varying effect on water exchange rates. While a sedentary person, such as an office worker, is losing only some five pints of water a day, a coal miner may lose 13 pints in one day's shift.

Even more extreme, a construction worker doing hard physical work under the hot sun, may have to replace over 20 pints of water on such a day. Dehydration and sweating can each cause a great drop in the excretion of urine. Tropical conditions can cause urine loss to be less than a pint of urine a day, often much less. On occasion, perspiration loss in the heat can be as much as three and one-half pints an hour.

Even under normal sedentary conditions in a temperate climate, there is a rapid turnover in the body's water supply. A 156-pound man possesses 70 to 80 pints of water within his frame. Half of it will be lost and replaced within 10 days. The rest of it will be recycled over and over again. Eventually, all the water in his body at any given point will be excreted and replaced.

WHY DON'T MOST PEOPLE DRINK MORE WATER?

Occasionally you may hear someone say, "I just don't drink much water." What they probably mean is they don't drink much plain or tap water. Practically everyone would benefit by drinking more water, but only if it's uncontaminated and unadulterated. Not more sugar-laden or diet colas, tea or coffee (with or without caffeine), artificial sweeteners such as aspartame and artificial creamers. Not more beer or other alcoholic beverages. When you think about it, preparing and serving these beverages is actually a form of deliberate water contamination, isn't it?

Why do you suppose so many people still turn to these unhealthful drinks over plain water? Could it be they simply don't like the taste of their tap water? Not exciting enough for them? Perhaps they don't feel like they're being a good host or hostess unless they've added something to the water? Or maybe unconsciously, they just don't trust the quality of their tap water. Whatever their reasons, what their bodies really need is more unadulterated pure water and I'll explain why….

ALL SODAS ARE SYNTHETIC

Most soft drinks are strictly a chemical product. When sugar is not used in their preparation, but a synthetic sweetener is, the manufacturers of such products advertise that the drink contains "less than one calorie!" This is an outright admission these soft drinks indeed have no food value, just man-made chemicals. Beer and other alcoholic drinks do have calories, along with addictive chemicals. Furthermore, such beverages act as diuretics as they draw water from cells and cause an increase in the flow of urine. Again, good pure water should be our beverage of choice.

> *SOME PRACTICAL ADVICE: If you feel you must occasionally indulge in coffee or tea, make it with distilled water. Most people enjoy the taste much more. In addition, you'll need considerably less coffee grounds or tea leaves, so you'll save some money there. More importantly, you won't be getting the uninvited chemicals. They are usually dissolved unbeknownst to us, in the untreated tap water routinely used to reconstitute soft drinks or to produce alcoholic and other beverages.*

While fruit juices such as orange juice, tomato juice, cranberry juice, apple juice, and pineapple juice are usually considered a good source of water, it is common today for such juices to be reconstituted from a concentrate. Unfortunately they frequently contain untreated tap water from the local area where that product was bottled or packaged! If you're lucky, the water may have at least been cold-filtered but that's probably about it. Read those labels! It just may surprise you to find out that you're actually consuming tap water from some very undesirable sources!

RECONSTITUTE JUICES AND BABY FORMULA WITH DISTILLED WATER

Even though fruit juices should not be considered a substitute for pure water, it is much healthier to buy the concentrate and reconstitute it yourself with your own pure distilled water. It's cheaper too! Also, if you have a baby in the family, be sure to make his or her formula with distilled water. You will most likely want to use slightly less powdered or liquid formula, especially for newborns.

A number of potentially toxic metals have been reported in breast milk as well as in baby formula made with tap water. Metals such as lead, mercury and cadmium do not occur naturally in human bodies. It is most likely that these metals get into our bodies by something we eat, drink or inhale.

Expectant mothers are subject to these contaminants. As a result infants are likely to be exposed to higher levels of these metals before birth than in breast milk or formulas made with contaminated water. How much better it would be for the mother and her newborn, if she drank distilled water during pregnancy and while breastfeeding and mixed any baby formula with distilled water.

> *Nitrates, which are now commonly found in tap water in many areas, are also a particular threat to babies under six months.*

In the last 40 years, there has been more than a 12-fold increase in the use of synthetic nitrogen fertilizers. Naturally, this has led to higher, and in some cases, dangerous levels of nitrates in food grown on such soil. In addition, much higher levels of nitrates are appearing in the run-off water and in our wells. The EPA limit for nitrates is 10 parts per million of nitrate as nitrogen.

In the presence of certain anaerobic (oxygen-demanding) intestinal bacteria, nitrates (NO3-) are converted to nitrites (N02-). (The bacteria are simply stealing one oxygen atom from the nitrate ion.) It is actually the nitrites that are particularly harmful to young infants, pregnant women, the elderly and anyone else with a low resistance to disease.

Nitrites combine with hemoglobin to form methemo-globin which can't adequately transport oxygen to body tissues. In infants, this is referred to as the blue baby syndrome. For pregnant women, the presence of nitrites in the blood stream can lead to birth defects, especially during the first trimester of pregnancy when the baby's vital organs are being formed.

So again, make baby formula with safe, distilled water. As stated earlier, you will probably need to cut back a bit on the amount of formula powder you use to make a batch. Most importantly, your innocent baby may escape the possible harm from some chemicals you least expect to be there. You'll both sleep better too.

Chapter 4:

The Chemistry of Water

Water is extremely stable. It carries many chemicals, either in suspension or in solution, without being permanently changed itself. Given enough time, water will dissolve at least a little of everything with which it comes in contact. As you will soon learn, this particular characteristic of water accounts for its almost limitless uses and also many of its problems.

> NOTE: Readers' discretion is again advised. In the next few sections, we get into a little basic chemistry, but don't worry, it's pretty simple. If you must, skip reading any section marked with an asterisk.

CHEMICAL FORMULA FOR WATER*

Water's chemistry appears to be quite simple, and it actually is, at least compared to most other chemical compounds. Most of us know water as a molecule consisting of one atom of oxygen (the most abundant element in the earth's crust) and two atoms of hydrogen (the most abundant element in the universe).

Scientists have taught us to express this chemically as H_2O, which is the empirical formula for water by volume. More simply put, two cubic feet of gaseous hydrogen will combine with one cubic foot of gaseous oxygen to produce two cubic feet of water vapor. This could just as well be stated as any other two-to-one ratio of volumes of hydrogen and oxygen.

Water is easily produced when hydrogen meets oxygen in the presence of a spark. In other words, to "make" water, hydrogen must be oxidized or burned. (It is not nearly so easily parted.) It is the way in which the hydrogen and oxygen combine to form the water molecule however, that gives water many of its special properties.

In reality, H_2O is only the formula for water in its gaseous phase. Liquid water is at least two or more, and usually a lot more, molecules of H_2O linked together with weak hydrogen bonds. The number of H_2O molecules linked together is directly dependent on the temperature of the water. The longer the strand of H_2O molecules, the colder the water is and vice versa. Liquid water is formed when water vapor is compressed to 1/1642 of its original volume. Only in its solid form (ice or snow), are all of the water molecules linked together by hydrogen bonds.

WHY DO SNOWFLAKES HAVE SIX POINTS?

A snowflake's six-pointed star is one modification of a more common form taken by ice crystals. The bond angle between the hydrogen atoms in a water molecule (usually about 104.5°), is very close to the interior angles of a hexagon (105°). This is the reason that when a snowflake sublimes from cool water vapor, it tends to take the form of a six-pointed crystal. The fact that the angle between the hydrogen atoms of a water molecule is not exactly 105° and will shift and vary under internal stresses as it grows outward in the process of forming its crystal, accounts for each snowflake being different.

WHY DOES ICE FLOAT? *

While the position of the hydrogen atoms on the oxygen atom gives the water molecule its polarity, it also accounts for another very unusual characteristic of water. As was explained earlier, most substances are denser, or heavier, as solids than as liquids. Water on the other hand, due to its polarity (explained in next section) and the particular way in which the molecules line up to form the crystalline structure of ice, expands 9% and thus becomes lighter as it freezes. This is why ice floats in water. Remember this when pondering an iceberg. Ninety-one percent of the iceberg is under the water surface!

IN H₂O MOLECULES, THE HYDROGEN AND OXYGEN ATOMS ARE SHARING ELECTRONS. *

Below is a simplified chemical diagram of a single water molecule and a diagram of several water molecules coming together to form liquid water through hydrogen bonding. Because the hydrogen and oxygen connect by sharing electrons, the chemical bond between them is extremely tight. Second, because of the top-heavy position of the hydrogen atoms in the molecule, the molecule is a polar particle and acts somewhat like a tiny bar magnet. The top-heavy hydrogen area is positively charged and the bottom area is negatively charged. These two features of the water molecule, its tight bond and its polarity, have a considerable effect on its behavior.

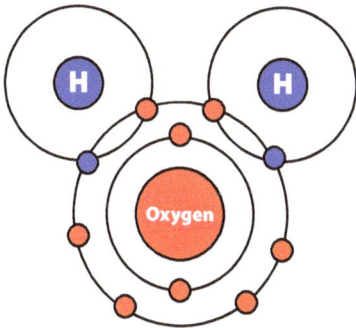

One water molecule formula exists in gaseous phase.

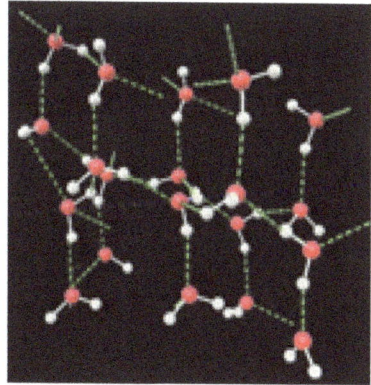

Positive and negative charged ions act like magnets, creating a tight chemical bond.

Because of its strong connective structure, water possesses amazing tensile strength. When enclosed in a slender, air-tight tube for instance, water can withstand a pull of 5,000 pounds per square inch. That is as strong as many metals!

The bonded strength of the individual water molecules also prevents them from ordinarily reacting with one another in a pure state to produce ions, or charged particles. What this means is that a water molecule, H-O-H, does not readily dissociate, or break apart into particles of H+ (a positively charged particle H+ with acidic properties) and OH- (a negatively charged particle with alkaline properties.)

WHAT IS PH?*

In nature this only occurs one time in every 10,000,000 molecules of water. Written in another way, that is 1/10,000,000 or 1/10 to the seventh power. This is where 7.0 as the pH of pure water comes from. Although we rarely discuss it, the corresponding pOH (concentration of hydroxide ions) of pure water is also 7.0. If one molecule of water in 1,000,000 molecules disassociates, or 1/10 to the sixth powers, that would be a pH of 6.0 and the pOH would be 8.0 and so on. The sum of the OH and the pOH is always 14.0. When you know one of them, you simply subtract it from 14.0 for the other. Hopefully the diagrams will make more sense of this for you.

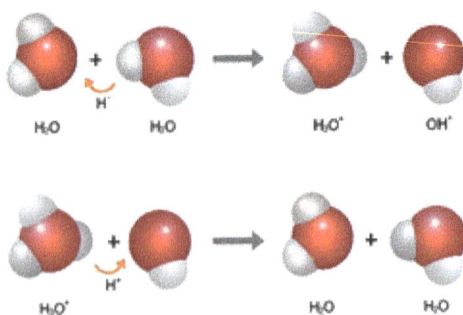

This particle H+ immediately attaches to another water molecule, H-O-H, to form H3O+. The remainder of the original water molecule is OH- which is a negatively charged particle with alkaline properties. (See the top row in the diagram.)

The two particles H3O+ and OH- then almost instantly reunite to form two molecules of water. (See the bottom row in the diagram.)

Absolutely pure water (which doesn't actually exist) would register 7.0 on a standard acid-base scale. Distilled water that has been exposed to air usually has at least a little carbon dioxide from the air dissolved in it. This results in a weak carbonic acid. That is why the pH of distilled water will frequently test very slightly below seven.

←	Acidic	—	Neutral	—	Basic	—	→		
0 1	2 3	4 5	6 7	8 9	10 11	12 13	14		

Battery Acid — Lemon Acid — Wine — Normal Rain — Distilled Water — Baking Soda — Soft Soap — Ammonia — Lye

The pH of distilled water is basically meaningless since distilled water has virtually nothing in it to buffer. When distilled water with a pH of 7.0 is required for laboratory uses and other specialized applications, there are collection procedures which can be used to produce carbon dioxide free water.

For consumptive uses, these procedures are unnecessary. In fact, when stored in a glass container, distilled water, with a pH slightly below 7.0, will taste even better and will exhibit a slightly longer biological shelf-life. Bacteria do not do well in liquids with a pH below seven, and in fact, rarely survive in a solution with a pH below 5.0. That is why vinegar is so widely used in food preparation and canning.

There are three main factors that account for a slight range of pHs for distilled water: (1) the length of time the distilled water is in contact with air, (2) the temperature of the distilled water at the time of the pH test, and (3) the concentration of bicarbonates in the water that was distilled.

Since the atmosphere always contains a small percentage of carbon dioxide, as stated earlier, the longer the distilled water is in contact with air, the more carbon dioxide will be dissolved in the distilled water…and thus a little lower pH reading.

There is a physical limit at any given temperature as to how much carbon dioxide can be absorbed by water. (The colder the water the more carbon dioxide—or any other gas as far as that goes—that can be dissolved in it.) Heating distilled water will drive out the carbon dioxide as well as any other gases. This will cause the pH of the distilled water to rise temporarily, but never above 7.0.

When the feed water containing bicarbonates enters the boiling chamber of a distiller, the bicarbonates decompose into carbon dioxide (there it is again), and carbonate.

Thus the pH will be lowered immediately since carbon dioxide is a non-volatile gas and not removed by distillation. There will be a temporary adsorption of the carbon dioxide when it passes through an activated carbon filter. This will cause a slight rise in the pH temporarily so again, this really does not mean much.

BOILING DISTILLED WATER RELEASES ANY CARBON DIOXIDE THAT MAY BE PRESENT

When distilled water is used for consumptive uses, the pH of the distilled water is of virtually no consequence. Heat will drive out the carbon dioxide during cooking.

NOTE: This is one of the reasons that soft drinks are heavily carbonated: it kills any bacteria and prevents them from multiplying. When sodas lose their carbonation (go flat) they are prone to bacterial infection from

> *the air. Also be aware that bacteria can live and grow in seltzer water...especially at certain pH levels.*

BACTERIA CAN ALSO THRIVE IN 'KANGEN' OR OTHER SO CALLED 'ALKALINE' WATERS

Bacteria not only survive in alkaline water, they actually thrive in water with a pH of eight to nine. So much for consuming alkaline water! It is NOT the healthful thing to do, mainly because alkaline water may have biological impurities as well as chemical impurities in it.

As you may or may not already know, pH is a logarithmic scale. This means that each digital difference of 1 is really a factor of 10. A digital difference of **2** is really a factor of 100, a digital difference of **3** is rally a factor of 1000, etc. Referring to the pH scale of many common substances, you will see that the pH of distilled water is much higher than most other common beverages which people frequently consume. Coffee, tea, fruit juices, and sodas typically measure in the range of 3.0 or lower on a pH scale. The pH of the gastric juices in our stomach is at or near 1.0! See this and the pH of other common liquids on the colored pH scale.

pH	Substance
14	
13	Bleach
12	Soapy water
11	Ammonia solution
10	Milk of magnesia
9	Baking soda
8	Sea water
7	Distilled water
6	Urine
5	Black coffe
4	Tomato juice
3	Orange juice
2	Lemon juice
1	Gastric acid
0	

Contrary to what the proponents of consuming alkaline water believe, the pH of our bodily fluids is largely controlled by our kidneys and lungs, and varies in different parts of the body, even within different regions in and around cells. There is no one pH for all uses in the body! Your kidneys regulate the pH in your body so there's no need to try and fight it.

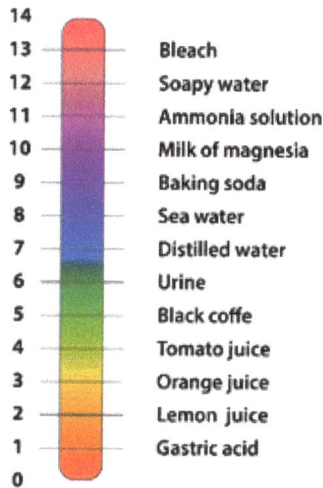

One of the reasons that uninformed potential distiller customers some-times give for not purchasing a water distiller has to do with the pH of distilled water. These unfounded objections have become more prev-alent as the number of companies that market alkaline water (usually water with a pH between 8.00 and 9.00 or higher) and other forms of Wonky Water have appeared.

THERE ARE MANY NAMES FOR ALKALINE WATER

Alkaline water is marketed under many different trade names, includ-ing *Alkazone, Alkapuro, Alka Way, Structured water, Kangen water* just to name a few. They are all listed on a very revealing website titled www.chem1.com/CQ. Check it out, in particular *"The Gallery of Water Related Pseudoscience."* This is one list where manufacturers of legitimate water products do not want be found! No distiller company is listed there.

Advocates of alkaline water claim all sorts of wonderful benefits, most likely because of the placebo effect or because they are simply drink-ing more water. There is no hard scientific evidence behind any of these unconfirmed improvements to their health, regardless of what enthu-siastic owners of these products initially claim. Unfortunately, the bad effects to an alkaline water consumer are rarely publicized. If you are near a computer, do a search for the dangers of alkaline water. I am sure your eyes will be opened.

Clever marketing has convinced many gullible consumers into thinking that it is the way to go as far as consumable water is concerned. There are at least three reasons why they are wrong. We will expand on them here:

1. The alkaline water that is produced still has impurities in it. Some of the impurities could have harmful immediate or long-term health effects. This means that, theoretically, every man-made alkaline wa-ter is likely to end up being at least slightly different, depending on

the tap water used to start out with. (In other words, there is no consistency for alkaline water as far as purity is concerned.)

2. Gastric juice is a digestive fluid formed in the stomach. It has a pH of 1.5 to 3.5 and is composed of mainly hydrochloric acid (HCl) around 0.5%, or 5000 parts per million. The pH of the stomach must be quite acidic to digest proteins, especially meat. When alkaline water with a pH of 8.0 to 9.0 is consumed, the pH in the stomach is temporarily raised. If meat and other proteins are in the stomach, poor digestion will most likely occur, often resulting in Gastro-Esophageal Reflux Disease. GERD is a condition in which the stomach contents (food or liquid) leak backwards from the stomach into the esophagus (the tube from the mouth to the stomach). This action will often irritate the esophagus, causing heartburn and other more severe and chronic symptoms.

3. The actual effect achieved is the exact opposite of what advocates claim it is! As was stated earlier, the pH varies throughout the body. It is even different inside and outside of our individual cells. It does not make sense to think that altering the pH in the stomach with liquids or solid foods will automatically adjust the pH of all the other liquid components or areas to exactly the pH needed. It is not that simple. The kidneys and lungs adjust the pH in various parts of the body primarily through responses of our circulatory and respiratory systems to keep pH at the proper level for our different bodily needs or functions. Elimination of carbon dioxide is one of the mechanisms used. Healthy bodies are always striving for balance. Alkaline water upsets this balance.

Adding alkaline food or liquid to the stomach will cause healthy kidneys to react by kicking in H+ ions (Hydrogen ions) to acidify the contents of the stomach, not alkalize the contents. Again, this is just the opposite of what advocates of alkaline believe that they are achieving. This creates an environment more favorable for stomach and colorectal cancer cells!

Those with kidney problems should be especially sure that they are drinking pure, distilled water, not alkaline water. Give the kidneys a rest.

When it is necessary or desirable to buffer distilled water, it will not take much of any alkaline substance to do this. Consumers could save thousands of dollars by alkalizing water themselves. In the home, the tiniest amount of sodium bicarbonate (baking soda) will do just fine. (Less than 1/16 of a teaspoon per gallon will do it.) Check with p-Hydrion test strips to get the pH you desire.

It is true that distilled water should never be plumbed through copper or aluminum as it will tend to dissolve light metals. This is partly due to the slightly reduced pH of the distilled water but more so due to the absence of dissolved solids in the distilled water.

OXIDATION REDUCTION POTENTIAL (ORP) AND YOUR HEALTH

Another argument the advocates of alkaline water have is that alkaline water has a more favorable (negative) ORP. What on earth is that you say? Let's explain it.

We do not see it (thank goodness) but there is a continuous exchange of electrons that takes place all around us between substances in the earth, in water, in the air and even in our bodies. Substances that are lacking electrons are desperately seeking to find and steal electrons from wherever they can find them. These substances are referred to as oxidizing agents.

Likewise there are substances that have extra electrons and are eager to donate their surplus electrons wherever they can. They are referred to as anti-oxidizing agents or reducing agents. Therefore electron exchange occurs in an attempt to reach electrical stability.

ORP stands for oxidation-reduction potential and indicates the degree to which a substance is capable of oxidizing or reducing another substance. It is measured by an ORP meter which is calibrated in millivolts (mv) (thousandths of a volt).

A positive ORP reading indicates that a substance is an oxidizing agent, (needs electrons). Oxygen is a very good example on an oxidizing agent…hence the name. The higher the reading the more oxidizing it is. For example, if a substance has an ORP of 300mv, it is three times more of an oxidizing agent than a substance that has an ORP of 100mv.

A negative ORP reading indicates that a substance is a reducing agent (has extra electrons). The lower the reading is (more negative) the more anti-oxidizing or reducing it is. As an example, if a substance has an ORP reading of -300mv, it is three times as much of a reducing or anti-oxidizing agent as a substance that has an ORP of -100mv.

Tap water and most other types of water, bottled as well as distilled, have ORPs that are positive indicating that they are oxidizing agents. (Alkaline water is a rare exception.) They have a fairly wide range of numerical values, measuring in the hundreds. These differences are rather significant when using them in a chemistry or physics laboratory. When using them for consumptive purposes, however, ORP readings alone really mean very little.

Again various organs of our body (especially our stomachs and small intestines in this case) come to our rescue…just as our kidneys and lungs do when it comes to pH. One can't use pH and ORP readings of substances in glass containers outside of our bodies, to predict what effects these substances will have once they enter our bodies. Alkaline water enthusiasts and other distilled water detractors need to recognize that the human body is not a hollow metal pipe. Our bodies are complex chemical factories, and various parts of them react in a multitude of wonderful ways with the substances which enter them, beginning in our mouth.

DISTILLED WATER IS A POOR CONDUCTOR OF ELECTRICITY

Contrary to popular belief, water itself does not conduct electricity. Rather, it is the dissolved inorganic compounds or impurities in water that make it a good conductor. In other words, a conductivity meter is a measure of the inorganic solids dissolved in water. Distilled water actually makes a good electrical insulator.

WHY DOES WATER DISSOLVE SO MANY SUBSTANCES? *

The polarity of the water molecule is also responsible for one of the most distinctive and important properties of water—its solubility. Water can dissolve, to some extent, virtually any inorganic and some organic substances with which it comes in contact. The positive portion of the water molecule simply attracts negative particles or the negative end of the other polar particles; and the negative portion attracts positive particles or the positive end of other polar particles. In addition, either end of the molecule will attract neutral particles. For example, when a substance such as Sodium Chloride (NaCl) dissolves in water, its particles break away from one another and cling instead to individual water molecules.

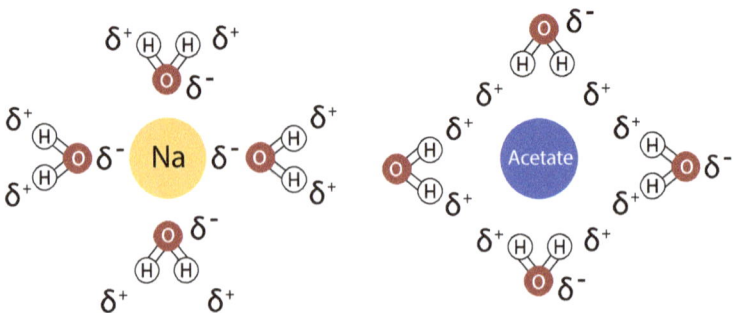

Not only does the hydrogen-oxygen configuration of the water molecule make it extremely attractive to other particles, it also makes it attractive to the appropriate end of other water molecules. It is the resulting intermolecular configuration that gives water a lower freezing point and higher boiling point than similar substances with non-polar molecules.

THERE ARE MORE THAN 64 FORMULAS FOR WATER

If water is not a source of fascination to most of us who take its behavior and its abundance for granted, it is nevertheless intriguing to chemists and physicists who continue to explore its structure and properties, and its occurrence in the universe. To date, more than 64 rare and different formulas for water have been identified, involving each of the various hydrogen and oxygen isotopes that have been identified, indicating we still have not learned nearly all there is to know about the chemical nature of this very common, yet unique and wonderful compound.

Chapter 5:

The History and Future of Water

WHERE DID WATER COME FROM?

In the beginning our earth was lifeless…that is until the water came. The first living things were water plants. The first animal life lived in water. Today we are still water animals, forced to live close to water, or perish.

The origin of water is as open to scientific speculation as the origin of the earth itself. What we do know is that water covers approximately three-fourths of the earth's surface—some 326 million cubic miles—and that 97% of it is in the oceans.

Do you realize if the earth were to be leveled off, the oceans and lakes filled in, water would completely cover the earth to a depth of nearly two miles? Yet if the earth were reduced to the size of a basketball and dipped into a bucket of water, the amount that would cling to the basketball would be in about the same proportion.

Almost 97% of the water on our planet is totally unfit for human consumption. It is the remaining 3% that concerns us most—the 3% available to us as fresh water. The salty 97% that may be home to most of the world's aquatic life is virtually unusable for most human purposes other than transportation, unless it is thoroughly desalinated.

DISTRIBUTION OF EARTH'S WATER

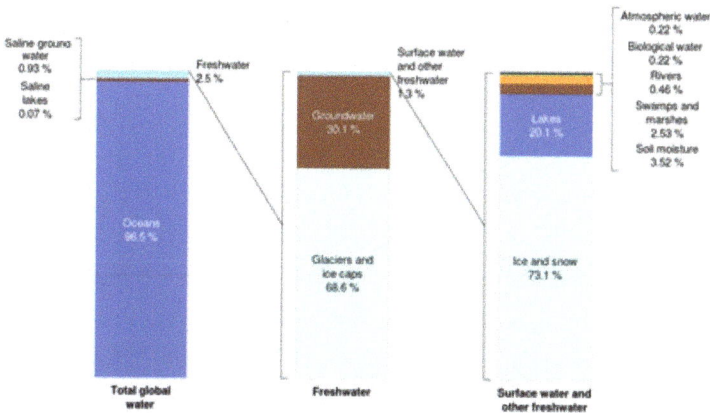

Saline ground water 0.93 %
Saline lakes 0.07 %
Freshwater 2.5 %
Oceans 96.5 %
Total global water

Surface water and other freshwater 1.3 %
Groundwater 30.1 %
Glaciers and ice caps 68.6 %
Freshwater

Lakes 20.1 %
Ice and snow 73.1 %
Surface water and other freshwater

Atmospheric water 0.22 %
Biological water 0.22 %
Rivers 0.46 %
Swamps and marshes 2.53 %
Soil moisture 3.52 %

This 3% fresh water still represents 33 trillion cubic feet, however. Unfortunately, three-fourths of it is frozen in glaciers and icecaps, with the remainder found underground, in lakes, rivers and in the atmosphere. Even at that, the accessible supplies of fresh water have served the earth sufficiently since it began.

FRESH DOES NOT MEAN CLEAN

Fresh water supplies aren't necessarily clean or uncontaminated and certainly aren't pure. The natural solubility of water, with its capacity to dissolve a myriad of substances with which it comes in contact, precludes the natural occurrence of water in an absolutely pure state.

The variety and concentration of dissolved substances in a particular water supply will vary depending on where the water has been. If, for instance, it flowed down a mountain side and into a lake, it may have come in contact with rock, decaying animal and plant materials, various gases, various metals, dust and dirt, micro-organisms, etc. it will have dissolved or absorbed at least a small portion of each. It might also carry in suspension other materials it cannot readily dissolve.

WHERE DOES IT GO?

Just as such contamination is natural to water, so is the hydrologic cycle —the natural purification process by which water can rid itself of impurities, at least temporarily, before picking them up again. The hydrologic cycle points out the indestructibility of water. Unlike many other substances, water cannot actually be consumed or exhausted. Rather, the earth possesses a fixed supply of water to which, through the hydrologic cycle, used quantities will always return. In short, water is reusable and recyclable. Every glass of water you drink contains water molecules that have been around since the earth began. In other words, it's water that has been cleansed repeatedly through the hydrologic cycle.

WHAT IS THIS HYDROLOGIC OR WATER CYCLE?

The hydrologic cycle is nothing more than an enormous distillation system powered by the sun. It goes into operation as the heat of the sun vaporizes the liquid water, breaks it down into individual H_2O molecules and draws it from the earth's surface into the atmosphere forming clouds, leaving virtually all of its impurities behind. As the vapor eventually cools in the clouds, it condenses and falls back to the earth as rain, snow or other forms of precipitation.

All of the water on our planet is in constant circulation and recirculation in what is known as the water cycle or more scientifically the hydrologic cycle. To me it is amazing that a drop of water can be trapped for thousands of years, frozen as ice. Then one day is melts and begins flowing toward the ocean. As the water in the ocean evaporates, it is carried by air currents, deeper and higher into our atmosphere. There it may become part of a cloud.

In due time, it sublimes on the surface of the smallest particle of dust to form a very small snowflake or condenses into a drop of liquid water. As it falls back toward earth, some of this water instantly re-evaporates. The rest of it continues to fall as a form of precipitation and when it reaches earth's surface, some of it immediately soaks into the ground. From the soil, plants absorb some of the water through their roots and stems and become a brief stopping off point in the water cycle.

The rest of it may again evaporate or drain into a surface supply such as a pond, lake, reservoir or stream. There the water will temporarily serve as home for frogs, fish, and a multitude of other aquatic creatures as well algae and other aquatic plants. Some of the water in the pond or lake may be lapped up by a deer or other land animal. The water in the deer is also a brief stopping off point in the water cycle.

The water cycle does more than cleanse the earth's water of contaminants. It also helps to moderate the earth's climate, making the equator cooler and the polar regions warmer than they'd otherwise be. Nor is the cycle synchronized around the earth. The rate of evaporation is greater in the warm temperatures of the tropics, for instance.

More significantly, not all of the earth's water moves through the cycle at the same rate. Much of the earth's fresh water is in underground aquifers. In fact, there is approximately 37 times more water underground than on the surface. At any given time, the greatest quantity of fresh (non-saline) water on earth exists as groundwater. Roughly 75% of the water used in the United States comes from underground sources. Underground water can be tied up for hundreds of thousands of years before it resurfaces again to take part in the hydrologic cycle. Some people are now even conjecturing that there may be a huge amount of water locked between the earth's mantle and its core. This idea is no longer just hypothetical as you will see from the next section. I believe you will find the following information to be quite fascinating.

A very interesting and detailed article relating to the earth's hydrologic cycle appeared in the June 16, 2014 *MSN Newsletter* (Microsoft News). It referred to a recent study that will help geologists and other scientists to further understand the earth's water cycle. According to this study, there are oceans' worth of water deep within the earth's rocky mantle, located between the earth's crust and the core of our planet. It is 400 miles below the earth's surface so we will have to depend on Mother Nature to get it to the surface of the earth. This water is not there as a liquid, ice or vapor state but bound by extreme pressure in a mineral called ringwoodite. Ringwoodite is a rare type of mineral that forms from olivine under extremely high pressure. It is trapped in the molecular structure as hydroxide ions (bonded hydrogen and oxygen atoms).

A geophysicist at the University of New Mexico, and the author of this recent paper in the Journal *Science,* gives scientists a new view of the structure of this part of the earth. This study also adds credence to the idea that the earth's water accumulated in the interior of the planet when the earth was formed, rather than arriving through the bombardment of icy comets, as some scientists have speculated. Direct evidence to support this new explanation has been lacking until now, so this is a great extension of our geologic knowledge.

According to researchers, this could lead to a better understanding of how plate tectonics move water between the surface of the planet and the interior reservoirs. This has been confirmed by laboratory experiments in which seismic waves deep in the earth were analyzed. The finding suggests that processes that occur in the shallower mantle that cause volcanoes and related activity at the surface, are also occurring farther down. Magma is created by a process called dehydration (melting) close to the surface of the mantle. This is responsible for volcanic hot spots all around the world. The increasing pressure causes minerals in the mantle to release their water, lowering the melting temperature.

Nature's recycling efforts are limited by the amount of water vapor the atmosphere can hold at any given time, approximately 3,100 cubic miles of it, or less than 100,000 cubic miles of the total water supply. As a result, the land areas of the earth receive only about 30,000 cubic miles of cleansed water each year from this continuous but slow cycle, with more than one-fourth of the total running off the land and into the oceans.

IT STARTS ALL OVER AGAIN AND AGAIN

The purification process begins with the evaporation of the surface water but in most cases gives way to re-contamination, just as soon as the vapor starts to condense and fall through the atmosphere. Once again, the solubility of water becomes operative, first attracting atmospheric pollutants and bacteria, then picking up salts, minerals and chemicals as it reaches the surface below. The cycle remains a marvel nevertheless, and while it involves only 0.005% of the world's water supply at any one time, the huge volumes of contaminated water cleansed and returned to us has, so far, been sufficient for our needs.

"That which the fountain sends forth returns again to the fountain."

— Henry Wadsworth Longfellow

Just as water moves throughout the earth, it also circulates in plants and in the bodies of all animals. "How does that happen" you may ask.

All living things are composed of cells. The smallest living things are made up of only <u>one cell</u>. One-celled, non-green plants is one definition for **bacteria**. Mono-cellular animals are referred to as protozoa. Larger plants and animals can be composed of billions of cells.

Cells give living things their structure and allow living things to perform all of the functions of life. The first of these functions is the circulation of nutrients. In animals this is referred to as digestion. Respiration is the circulation of oxygen and carbon-dioxide. (That is breathing for us animals.) Other important life functions include growth, reproduction and elimination of waste. These and any other life function can only be carried out in a watery environment; in simpler terms this means in the presence of water molecules. By now you probably know that water molecules are composed of two gases in the volume ratio of two parts hydrogen to one part oxygen.

WILL WE EVER RUN OUT OF WATER?

While the total amount of water on our planet is essentially a finite and fixed amount, it may be considered to be great by some standards. When one takes a closer look however, there really is not an overabundance of water, especially when we limit our search to water that is fit for human consumption. As we saw earlier, the total amount of water on earth is the same as it ever was. Remember the analogy of the film of water left on a basketball dipped in a bucket of water? That's not much for its many uses and the growing needs as the human population increases. The only way we will have enough water to go around is if we increase the amount of water we make available by recycling previously used water or salt water, using man-made technology. We will have no choice as there is no new water to be had!

HOW DO WE USE FRESH WATER?

Let's look at the uses for fresh water in four main categories:, domestic use, industrial use, business use and agricultural use.

DOMESTIC USE: Household use for water accounts for roughly 8% of our daily use. Most people are familiar enough with the need for water in their own homes, for drinking and cooking, for cleaning, for laundry, for bathing and for the toilet. Of these domestic uses, the water used strictly for drinking and cooking represents only 0.5% of the total. Examples, even with efforts to conserve water, include:

Typical bath	30 to 50 gallons
Shower	15 to 30 gallons
Laundry Load	18 to 30 gallons
Dishwasher Load	25 to 40 gallons
Toilet Flush	1.5 to 3 gallons.

The average American also uses another 100 gallons outside the home for such purposes as sprinkling lawns and gardens, washing the car or washing down sidewalks, and so on.

US HOUSEHOLD WATER USE

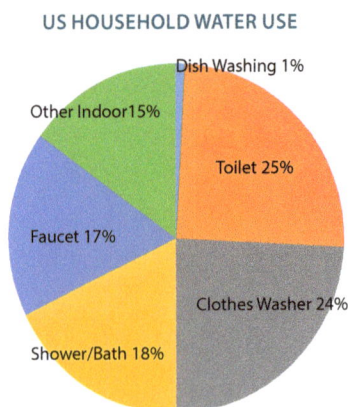

Dish Washing 1%
Other Indoor15%
Toilet 25%
Faucet 17%
Clothes Washer 24%
Shower/Bath 18%

Drinking water is only 1% to 2% of our total water use, but around the world, water consumption per inhabitant varies considerably. For example, the average American uses 132 gallons per day (500 liters/day), whereas the average western European uses only 40 gallons per day (150 liters/day) and the average African uses about only 13 gallons per day (50 liters/day). No one is equal in the amount of water used, even if these differences tend to decrease.

Another interesting difference is way of life and where people live. For example, people in the country use less water than those in the city. Also the age of people makes a difference. Young people use more water than old people, on the average. (Teenagers tend to take longer showers!) There are many factors to explain this discrepancy. The most important factor is the amount of water that is available. For example, because of its rarity, water is a very precious resource in Africa.

This ties in with another factor… price. Wealthier people consume more water. This is because they have bigger homes with more showers, bathrooms and other water-using appliances.

INDUSTRIAL USE: The domestic use of water is literally a drop in the bucket compared to its industrial use. Nearly 43% of the water consumed nationwide is used by industry. Water is the one common requirement of virtually all manufacturing processes and is utilized by industry in greater quantities than any other raw material. For instance, it takes 100,000 gallons of water to manufacture one automobile. To print one copy of a Sunday newspaper takes 280 gallons of water. It takes five full gallons of water (and a cow) to produce one gallon of milk.

BUSINESS USE: Water usage in the business sector is equally noteworthy. To operate a Laundromat with 10 washers requires 1,800 gallons of water a day. A carwash handling 24 cars per hour uses 8,000 gallons daily. This laps over into recreational and environmental use which is a small but growing percentage of the fresh water being used.

AGRICULTURAL USE: Even business and industrial consumption of water does not equal the amount used in agriculture. It is estimated that

GLOBAL WATER WITHDRAWALS

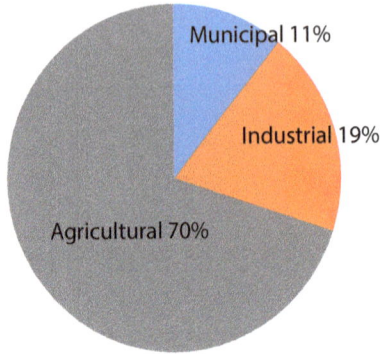

Municipal 11%

Industrial 19%

Agricultural 70%

2,000 to 3,000 quarts of fresh water are needed daily to take care of our dietary needs. A full 47% of our nation's water supplies goes into the production of food, not counting the amount used in processing and transporting food products. Believe it or not, it takes up to seven gallons of water to produce one head of cabbage, 85 gallons for a pound of wheat, 500 gallons for a pound of rice and 1,300 gallons for a pound of cotton. Livestock also requires a huge amount of water. While sheep can survive on one and one-half gallons of water per day, cattle on the average need eight to ten gallons per day, and horses as much as 15 gallons.

HOW MUCH DO WE USE IN TOTAL?

The fact that water is available on earth in a fixed, recyclable supply is both good and bad. Unlike many of our energy sources which exist in

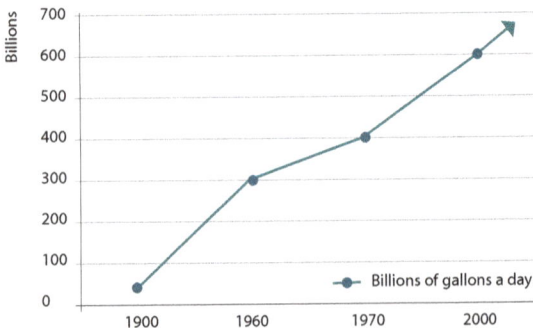

US water use since 1900.

finite, exhaustible supplies, water will always be with us. We have as much today as our ancestors had and we will have as much tomorrow as we have today. However, what was enough water for our ancestors, and perhaps enough even now, may not be enough for our children's generation. There are more users of water today than yesterday, simply due to population increases. Given our continued social, industrial and technological advances, there are new uses for water all the time.

The United States' supply of fresh water dependably provides us with between 400 and 500 billion gallons of water daily. In 1900, the average daily usage was 40 billion gallons. In 1960, we used a daily average of nearly 300 billion gallons. And by 1970, we were using nearly 400 billion gallons daily. Today our needs are beginning to exceed our supply! That is why more widespread, man-initiated recycling will be required.

WATER IS NOT EVENLY DISTRIBUTED

Our water supplies are not distributed evenly around the country. With increased usage, the availability of water in the Southwest part of the United States, where it never has been plentiful, could seriously threaten the future development of the area. Even in the Northeast, where fresh water supplies are far more abundant, the demand for water by the huge population and industrial centers has necessitated the reuse of water before it has time to enter the hydrologic cycle for cleansing. In effect, we have to supplement Nature's purification process with processes of our own.

WHY THE WORLD'S WATER PROBLEMS WILL NOT BE SOLVED ON THEIR OWN

Even though water provides no calories or organic nutrients, safe drinking water is essential to humans and all other forms of life. Access to safe drinking water has improved over the last few decades in almost every part of the world. According to the World Health Organization, more than one billion people in low- and middle-income countries lack access to safe water for drinking, personal hygiene and domestic use. These figures represent more than 20% of the world's people. In addition, close to three billion people do not have adequate sanitation facilities.[5]

Water plays such an important role in the world economy, as it functions as a solvent for a wide variety of chemical substances and facilitates industrial cooling and transportation. Today, approximately 70% of the fresh water used by humans goes to agriculture. This trend will no doubt continue to increase as the world's population and the need for food increases.

Population growth is accompanied by the increased need for the use of more and more chemicals of all kinds. Starting with personal care products and pharmaceuticals, add to this solvents and virtually all other chemicals in everyday use and we have a growing disposal problem and the risk of additional contamination.

---

5   World Health Organization/UNICEF Joint Monitoring Task Force, wssinfo. org

Chapter 6:

Dangers to and from our Water Supply

n virtually all parts of the country, sanitation procedures are necessary to rid public drinking water of pollutants. Even the simplest natural pollutants, mineral deposits, bacteria and parasites such as giardia or Cryptosporidium can invade a water supply, rendering the water either unappetizing or actually hazardous for drinking. There does appear to be a predominance of widespread water problems in third world countries when it comes to those of micro-biological origin. Cholera is a prime example of this.

Cholera germs can be picked up from unsafe drinking water or from eating oysters that have ingested cholera-carrying zooplankton. The severity of the diarrhea and vomiting can lead to rapid dehydration and electrolyte imbalance, and even death.

The recent cholera outbreak in Haiti, for example, is the worst epidemic of cholera in modern history, according to the US Centers for Disease Control and Prevention.

After the 2010 earthquake, in little more than three years, as of August 2014, it has killed more than 8,231 Haitians and hospitalized hundreds of thousands more while spreading to neighboring countries, including the Dominican Republic and Cuba. Since the outbreak began in October 2010, more than 6% of Haitians have had the disease. Worldwide as of 2012, it affects three to five million people and causes 100,000 to 130,000 deaths each year. On a side note, cholera was one of the earliest infections to be studied by epidemiological methods.

WATER-BORNE ILLNESSES ARE NOT JUST A PROBLEM IN THIRD WORLD COUNTRIES

We tell ourselves, when it crosses our minds, that in the United States our water is basically safe to drink. We may attempt to reassure ourselves it is only in third world countries, or perhaps near a few hazardous waste dumps in the United States, that drinking water is actually unsafe. This is, however, not the case as you will see in the following examples.

Many of you probably remember the widespread sickness in Milwaukee, Wisconsin a little more than 20 years ago, caused by a protozoan cyst called *Cryptosporidium*. Newspapers at the time documented that more than 400,000 people of all ages became violently ill, several thousand were hospitalized and over 100 died. This is one of the largest outbreaks of water-borne disease in the United States in recent times.

More recently, water contamination incidents in New York City and numerous small rural communities throughout the US caused thousands to boil their water. With all the information now available, you would think that the safety of our drinking water supply would be well assured. Unfortunately this is not the case. With the introduction of humans or animals into an area, bacterial and protozoan contamination of water supplies invariably increases because the human and animal waste, as well as decaying garbage, eventually find their way into our water supplies.

AGRICULTURE IS ONE OF THE MAJOR POLLUTERS

As alluded to earlier, agriculture is one of the major users of water, and is also one of its major polluters. In addition to decaying plants and animals; fertilizers, pesticides, herbicides and fungicides can be carried off by rain to the nearest surface water supply or can seep into the soil. Eventually, they even contaminate the underground aquifers (well sources). Naturally, most of the water problems that stem from chemical contaminants appear to be a greater problem in the agricultural areas than non-agricultural areas.

For example, the common agricultural herbicide atrazine, is toxic to wildlife and is known to impact the reproductive health of wildlife. It is applied to field crops by tractor-drawn equipment, by aerial spraying and by chemigation. Chemigation is a rather innovative way of applying a mixture of agricultural chemicals, using a center-pivot irrigation system. Although chemigation involves some additional risks to the environment, it is now a well regulated and monitored agricultural practice in many parts of the Midwest. Unfortunately, atrazine has contaminated watersheds and drinking water throughout much of the United States, according to a news report that was released in August of 2009 by the Natural Resources Defense Council (NRDC). This certainly raises concern about the chemical's effects on the health and reproduction of humans since we know how toxic it can be to animals.

This whole process speeds up immensely when the contaminated water seeps into abandoned wells that were not properly sealed. There are hundreds of thousands of these around the country, according to estimates from various water well associations. While the 2009 NRDC report indicated widespread contamination due to atrazine, contamination was most severe in the states of Illinois, Iowa, Indiana, Missouri, and Nebraska.

Atrazine has a half-life of several years (longer in colder climates), and can be detected in almost every stream and river of the US. Eventually, much of it gets to the Gulf of Mexico where it continues its plant-killing spree of algae and other beneficial water plants that provide food and oxygen for aquatic life.

Banned by the European Union, atrazine is regulated in the United States by the US EPA. Under the Safe Drinking Water Act, the EPA has determined that an annual average of no more than three parts per billion (ppb) of atrazine may be present in drinking water. Some scientists are concerned about exposure for children and pregnant women, as even small doses could impact development of the brain and reproductive organs. Atrazine also acts as a multiplier to increase the toxic effects of other chemicals in the environment. There is very strong evidence that atrazine is what they call an 'endocrine disrupting chemical' (EDC). EDCs interfere with critical reproductive hormones even at extremely low levels.

A 1994 report by the National Wildlife Federation details new evidence linking commonly used herbicides such as atrazine, alachlor, 2,4-D and insecticides such as chlordane and malathion to a variety of wildlife and human health problems. This report cites research, mostly completed since 1988, indicating a number of chemicals ranging from DDT to dioxins may be responsible for reproductive problems and endocrine (hormone) system disruption in both people and animals.

Those effects include increased rates of breast, prostate and testicular cancer, impaired immune systems, deformed reproductive organs, learning and behavioral problems in children, infertility and endometriosis. Most of these chemicals persist (are slow to break down) in the environment and have a tendency to accumulate in fatty tissues of animals so they are passed through the food chain.

According to this detailed report, recent studies confirm what has been suspected by some researchers for some time. PCBs, DDT and a host of other chemicals can act as hormone copycats and mimic the effect of hormones like estrogen and testosterone in humans and wildlife.[6]

This may account for a 50% drop in sperm counts of men in industrialized countries since 1930, breast cancer has doubled since the 1950s, and reproductive problems have increased for various species of wildlife, and for humans. Farmers and farm workers who use and apply such chemicals are among those at greatest risk.

There is currently a long overdue nationwide effort to revise the federal Clean Water Act to phase out or ban some 70 chemicals which can accumulate in living tissues. We should all urge the Environmental Protection

---

6   Dore Hollander, "Environmental Effects on Reproductive Health: The Endocrine Disruption Hypothesis," Family Planning Perspectives, Vol. 29, No. 2, March/ April, 1997.

Agency to consider the health effects of such chemicals when approving their use for agriculture.

INDUSTRY CREATES EVEN GREATER POLLUTION

By far the most serious water pollution problems can be attributed to modern industry. More than 30,000 industrial plants in the United States require large volumes of water for production purposes. Most of them are therefore situated near a large body of water for easy access. However, the close proximity of these water supplies to the industries they serve also makes a convenient repository for the waste material most industries must discharge as a result of manufacturing processes. In fact, virtually all public and industrial sewage systems are designed to eventually empty into a water supply.

SAFE DRINKING WATER ACT AND EPA EMERGING CHEMICALS FOR REGULATION

In 1974 Congress passed the Safe Drinking Water Act (SDWA) to provide national requirements for drinking water standards to protect the public health. This law was tightened considerably in 1986 and again in 1996 by subsequent congressional action. This is an ongoing process but many additions and changes are now long overdue. A serious nationwide effort is needed to revise the National Safe Drinking Water Act again because of whole new categories of contaminants. (The Federal Clean Water Act is seriously in need of updating for the same reason.) The Environmental Protection Agency (EPA) continues to research the health effects of various contaminants suspected of adverse health effects.

This work is slow and arduous and it takes many years of research before a contaminant is added to the primary contaminant list. Enforcement of the SDWA was assigned to the EPA, which currently sets maximum allowable limits or health based standards for around 100 specific contaminants. There are four main categories of contaminants on the primary contaminants list, all known to have direct and sometimes immediate health effects. They are: inorganic chemicals (natural), organic chemicals (carbon-based, both natural and synthetic), microbiological contaminants (bacteria, protozoa, viruses, etc.), and radiological contaminants (natural and synthetic).

The EPA also sets standards for the "secondary contaminants" list that consists of contaminants that may cause cosmetic or aesthetic effects such as smell and color. While the primary contaminant list is enforceable with daily fines, the secondary contaminant list is voluntary and non-enforceable. Because we can more readily sense their presence, we are sometimes more concerned about these secondary contaminants than we are about the primary contaminants, although we shouldn't be.

Considering that there are many more than 85,000 known chemicals in daily use, the number that are being regulated seems to be woefully inadequate. But based on the SDWA, water that meets the standards set by the EPA is deemed to be potable or safe. This generally means that a glass of this water will not kill you but it says very little about the long range health effects. Sometimes a level that is deemed to be "safe" at some point in time is found to be too high later and the limits are tightened to lower levels as "acceptable" for public health. This is a good thing. For this reason, many believe that no level of contaminant is acceptable if it can be eliminated from drinking water. When distilling your own, water that is exactly what is happening. Distillation systems remove virtually all contaminants from the water. Third-party testing by independent laboratories has confirmed this.

New chemicals are continually being produced and at an alarming rate. You can bet that many of them will eventually end up in our drinking water. Because of this, a list of emerging chemical contaminants has been developed. It contains a multitude of harmful chemicals which are well known to accumulate in living tissues. This no doubt will result in a host of serious health issues and problems. We should all urge the Environmental Protection Agency to consider an immediate ban or at least a gradual phase out of many of these chemicals because of known serious health effects on humans, plants and animals.

Almost daily chemicals are being discovered in water that had not previously been detected, or are now being detected at levels that may be significantly different than previously expected. They are now being referred to as "contaminants of emerging concern" (CECs). Even though there is a risk to human health, and the environment associated with their presence, the frequency of occurrence is not well established and the source is often not known.

The Drinking Water Division of the EPA is working continually to improve its understanding of a number of CECs, particularly pharmaceuticals and personal care products (PPCPs) and per-fluorinated compounds, among others. There are two very valuable computer data bases that you should be fully aware of and refer to often for more information on the prevalence and harmful effects of various water contaminants. They should be referred to periodically because of timely changes in them.

The **National Primary Drinking Water Regulations** (http://water.epa. gov/drink/contaminants/index.cfm), which also includes Secondary Regulations, will inform you about current drinking water standards as well as known or suspected heath concerns. They are the main components of the Safe Drinking Water Act. This listing not only tells you what the current EPA standard or limit is but also what the health concerns are for each contaminant.

Water municipalities and any other public water supplies are subject to daily fines for not meeting these standards. Secondary standards are not federally enforceable.

The **_National Tap Water Data Base_** (http://www.ewg.org/tap-water/) tells you the exact known problem not only nationally, but also regionally, by state and even down to the individual zip code. While the vast majority of people living in the US drink water that meets the standards established by the EPA, there are millions who are drinking water from small wells that are not regulated. Even though the SDWA does not regulate private wells, meaning wells that serve fewer than 25 individuals, it does require multiple actions to protect drinking water and the sources of drinking water including rivers, lakes, streams and reservoirs. Many of these sources, especially in rural areas, contain bacteria, pesticides, herbicides and fungicides.

A LOT CAN HAPPEN TO WATER ON ITS WAY TO YOUR TAP

Pollution also occurs closer to our taps than we might realize. Even if it were possible to have the water free of all contaminants when it left the well, reservoir or local treatment plant, there would still be a question as to whether it was fit to drink by the time it reached your tap.

Mineral buildup in a tap water line from Indiana.

Since water is such a scavenger (sometimes described as hungry), the quality of our tap water is dependent upon the material used for the pipes and joints. Remember that the underground network of pipes consists of a wide variety of materials, old and new lead, galvanized iron, asbestos, re-enforced cement, cadmium, copper, brass and PVC. The same goes for the plumbing that carries water into your house and up to the kitchen sink.

Asbestos-contaminated water has been linked to higher incidences of certain types of cancer. An even more widespread and hazardous problem is lead. Based on new evidence of extreme toxicity, the federal limits for lead were recently dropped from 50 parts per billion (ppb) to five ppb at the source. Lead piping was commonly installed before the early 1900s and is still in widespread use even though its use for new installations was outlawed several years ago. Considerable amounts of lead can be picked up as water flows through the plumbing network, so the federal limits for lead at your tap are 15 ppb. Hundreds of community water supplies now exceed these limits. Zero ppb of lead in our drinking water should be our goal!

PLUMBING IS ALMOST BEYOND SOLUTION

There are cross connections of sewage drains with tap water lines in most cities. Most plumbing connections were made decades ago, long before the average plumber was aware of the danger of water contamination.

In many areas, but especially rural areas, sewage lines are located directly above the water delivery lines. Gravity causes cross-contamination. Worn and outdated water mains allow liquid contaminants, such as feedlot runoff, to seep into the water distribution network, carrying various disease-causing micro-organisms along the way.

MICROBIOLOGICAL CONTAMINATION MAY BE THE SCARIEST

Of the many hazards of water pollution, historically the most published has always been the numerous microscopic infectious agents. They appear as anaerobic (not dependent on free oxygen), bacteria, protozoa and viruses. Water can transmit a variety of diseases when it contains pathogens. Contagious diseases like cholera and infectious hepatitis were once rampant. Today, these diseases are largely controlled in the United States through the widespread use of chlorine or chlorine-based disinfectants in public water supplies and some rural supplies. Some larger communities are now turning to ozone (a concentrated form of oxygen) for that same purpose. It is quicker and more effective, but also more expensive and has little residual power since it quickly reverts back to oxygen.

Tap water from public water supplies is also generally free of dangerous concentrations of other micro-organisms because they are usually killed (but not always) when chlorine is used in water treatment. Yet, chlorine is not the cure-all it was once thought to be. Some protozoa can adapt and escape death by chlorine, because of their special ability to form cysts. Encased in the cysts, they are often protected from the destructive power of chlorine. This is exactly what happened in the 1993 episode with *Cryptosporidium* in Milwaukee.

VIRUSES ARE ABUNDANT IN MOST WATER

Viruses (actually fragments of DNA) can be up to 1,000 times smaller than bacteria (single-celled non-green plants) or protozoa (single-celled animals). They are therefore much harder to detect and are very common in water. A single teaspoon of lake water can contain over 1,000,000,000 (one billion, with a B) live viruses!

Most viruses can be killed with the addition of chlorine to the water. It is difficult however, to determine how many may still remain alive. It is thought that most viruses are harmless to humans, although there are significant exceptions such as the polio virus and the AIDS virus. Many viruses are thought to be responsible for flu and cold epidemics. There is also a fear among some scientists that the polio virus may have adapted to chlorine and become immune to its effects.

Infection can still be spread by direct contact with biologically contaminated water, for example, while swimming or bathing. In many parts of the world there are still areas where diseases such as polio are still quite prevalent through the drinking water, especially in times of war or natural disaster. Rwanda, Africa is a recent tragic example.

Viruses contaminate water as a result of human, animal and vegetable sewage deposits, and even careless disposal of medical waste. This class of contaminants still grabs the biggest headlines in today's newspapers and will no doubt continue to do so in the foreseeable future at an even faster pace. For example, the Ebola virus was recently the main topic for discussion on talk radio and TV and is the main news item on almost every news source. While air and water are not currently being considered possible routes for infection, they can't be completely ruled out at this time. Immunologists warn us that viruses have the ability to mutate, and could change into something totally different in just a couple of generations.

MANY INORGANIC SUBSTANCES ARE HARMFUL

Certain inorganic substances which are found naturally in some water supplies, in plumbing materials, or as a result of industrial wastes, can be harmful to the body in even minute amounts. These include such known poisons as arsenic, chromium, lead and mercury. To some extent, even copper and aluminum are toxic metals.

Compounds containing calcium, magnesium and iron are among the more common and plentiful inorganic substances in water and generally account for its hardness. Most of us are familiar with the hard water deposits found at the bottom of a tea kettle or around sink faucets and drains. We are also familiar with the many problems they can cause in our plumbing, hot water heaters and other appliances that use water. Perhaps you are still experiencing the difficulties these scaly substances cause in washing dishes and clothes, as well as bathing your own body. Nutritionists now speculate these inorganic minerals are very poorly assimilated from water. In fact, many doctors believe minerals from water may tend to accumulate in the body. Perhaps they're even related to certain gallstone and kidney stone problems, certain types of arthritis, hardening of the arteries and other degenerative conditions.

WATER TREATMENT ITSELF ADDS AN UNKNOWN AND PERHAPS A POTENTIALLY DANGEROUS FACTOR

In addition to the multitude of contaminants already in the water, a host of new inorganic chemicals are added as part of the water treatment process. Modern municipal water treatment companies use as many as 47 different chemicals to clean up the water. It is quite unlikely that all 47 of them would be found in a single supply, but you can be reasonably sure a number of them will be found in almost any city-treated water you might happen to drink. The list of chemicals is almost staggering to me. Some of these chemicals mix and can potentially react with the chemical pollutants already present to form new compounds. You might find it interesting to find out which of these chemicals are being used to treat your water. The list is as follows:

Activated carbon	Activated silica	Aluminum ammonium sulfate	Sodium hypochlorite
Aluminum sulfate	Aluminum potassium sulfate	Aluminum chloride (solution)	Sodium thiosulfate
Alum (liquid)	Ammonia (anhydrous)	Ammonia (aqueous)	Tetra-sodium pyrophosphate
Ammonium silico fluoride	Ammonium sulfate	Bentonite	Sodium silicate
Bromine (liquid)	Calcium carbonate	Calcium hydroxide	Sulfur dioxide (gas)
Calcium hypochlorite	Calcium oxide	Calcium dioxide	Tri-sodium phosphate
Chlorinated copperas	Chlorinated lime	Chlorine (gas)	Sodium sulfite
Chlorine dioxide (gas)	Copper Sulfate	Disodium phosphate	Sulfuric acid (liquid)
Dolomitic hydrated lime	Dolomitic lime	Ferric chloride	Sodium fluoride
Ferric sulfate	Fluorosilicic acid	Hydrofluoric acid	Sodium hexa1-netaphosphate

Ozone (gas)	Sodium aluminate	Sodium bicarbonate	Sodium hydroxide
Sodium bisulfite	Sodium carbonate	Sodium chloride	

When chlorine was first introduced in 1913, the only water problem most people were aware of was bacteria. A small amount of chlorine was added to the water and it practically wiped out typhoid epidemics and other waterborne diseases. With this apparent panacea, complacency set in and as the water grew dirtier and dirtier, larger and larger doses of chlorine were required. The amount of chlorine used has increased to the point that some are now questioning if the treatment might be worse than the diseases. How could this be? Most people don't realize that they are practically drinking bleach!

WHAT'S THE PROBLEM WITH CHLORINE?

Chlorine interacts with the humic acid in decaying plants to form chloroform. Chloroform is classified as a THM (trihalomethane) and is considered to be a carcinogen (cancer-causing agent). Chlorine reacts with almost every other substance to form a carcinogen. According to the publication *Think Before You Drink* from the Natural Resources Defense Council (September 1993), THMs are associated with 10,700 or more bladder and rectal cancers per year in the United States alone. That's twice as many compared to people who die from fires, and more people than are killed by handguns. For more information on chlorine, watch our video published on YouTube explaining the dangers of chlorine, https://www.youtube.com/watch?v=VhFJxk82Pn4.

Because of this, the EPA has established a maximum contaminant level (MCL) of 100 ppb for THMs. It has since been lowered to 80 ppb and it will most likely be lowered to 50 ppb or less in the future. In spite of this, many water utilities seem to be unruffled by the fact their standard operating procedure puts a potential cancer-causing agent into every glass of water their customers drink. Other books, such as *Coronaries/*

Cholesterol/ Chlorine by Joseph M. Price, M.D. (1984), present startling evidence that links treated drinking water with coronary heart disease and stroke. This book is still as meaningful as the day it first came out in 1969 and can still be found.

The EPA limit for chlorine in drinking water was long overdue. It has only been established and enforced at 4.0 mg/L or parts per million since January of 2002. At that time, limits were also set for the first time for disinfection by-products.

NOTE: The concentration of chlorine that is recommended for shock disinfecting of a well is 25 ppm. To my knowledge there is still no federal limit for the concentration of chlorine in swimming pool water but I feel there should be. Today, a common practice in more and more municipal water treatment utilities is to combine ammonia with chlorine to create chloramines which are more stable but less effective in killing bacteria.

Chloramines are especially hard on fish and on anyone with kidney problems. Chloramines and chlorine are toxic and must be removed from water used in dialysis. For more detailed information on this problem, check it out on the web at Citizens Concerned about Chloramines and in the San Francisco Bay area by calling them at 408-227-5767.

FLUORIDATION IS A CONTROVERSIAL PRACTICE

A growing number of people also consider the fluoridation of public water supplies another form of deliberate contamination and legalized national medication. The practice of fluoridation started to become popular during the 1950s after some poorly corroborated studies concluded the natural presence of calcium fluoride in certain water supplies in Hereford, Texas and other parts of the country reduced tooth decay. Many today believe this was due to the presence of calcium...not fluoride.

Unfortunately, very few questioned the possible harm and side effects. One well-known fluoride opponent who did was Dr. John Yiamouyiannis, in a well-documented book titled *Fluoride: The Aging Factor* (third edition revised and published 1993). Another part of the controversy surrounding fluoridation involves the common use of sodium fluoride, rather than the naturally occurring calcium fluoride which exists in many parts of the earth. Many are again questioning this harmful practice.

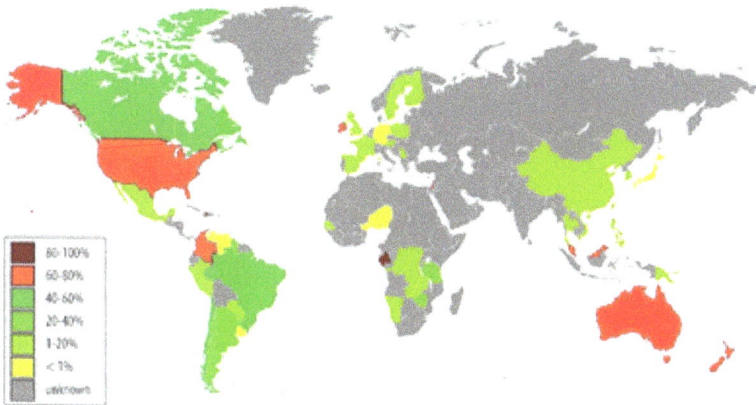

The map shows the present extent of fluoridated water usage around the world. Colors indicate the percentage of population in each country that receives fluoridated water, where the fluoridation is to levels recommended for preventing tooth decay. This includes both artificially and naturally fluoridated water as of March 2009.

ORGANIC CONTAMINANTS ARE NOW EVERYWHERE

The most numerous pollutants are the organic or carbon-based substances. They are discharged in the form of detergents, solvents, insecticides, pesticides or petroleum derivatives. These and other chlorinated hydrocarbons are chemical substances which do not dissolve in water, but which can readily interact with one another or with inorganic substances to form entirely new chemicals.

More than 20 million natural and synthetic organic compounds have now been cataloged. Because of the vast number of organic compounds being developed in labs annually, thousands of new substances are also introduced into the environment each year. The possible reaction combinations are multiplying astronomically. In most cases, neither the results nor the effects of these potential combinations have been determined.

Organic pollutants in water can have another serious effect. They attract decomposer bacteria, which use up the oxygen in water thus killing entire populations of fish. The shorelines of polluted lakes and rivers cease to be attractive to residential and resort developers and untreated pollutants can render water unusable for irrigation or other agricultural purposes.

HEAT CAN ALSO BE CONSIDERED A CONTAMINANT

The thermal pollution of water can also have adverse effects on the environment. Thermal pollution is actually nothing more than heated water that has been used as a coolant for certain industrial processes and put back into the water supply. An example is water that has been used to cool a steam-powered generator. Increasing the natural temperature of a water supply even a few degrees can disrupt vegetation and wildlife which are susceptible to changes in their habitat.

RADIOACTIVE CONTAMINANTS ARE LESS PREVALENT AND SOMETIMES HIDDEN BUT...

According to an article written August 1, 2014 by Sylvia Booth Hubbard, when many people think of radiation exposure, they quite often think of mushroom clouds following a nuclear explosion. She went on to point out that we are all exposed to radiation every day of our lives through the environment—the earth itself emits radiation—and through choices we make, such as using cell phones. According to the National Institutes of Health, the consequences of overexposure can be quite deadly and is believed to be the source of up to 10% of all invasive cancers.[7] A less prevalent, but potentially just as life-threatening form of water pollution is the presence of harmful radioactive substances such as radium, uranium, plutonium, iodine and cesium in the earth's crust. They can be introduced into a water supply as a result of mining and processing of radioactive material, faulty disposal or leaking of nuclear waste from power plants, even fallout from nuclear testing.

On March 11, 2011 a giant earthquake and tidal wave hit Fukushima, Japan. Initially the world watched in horror at the amazing destruction of the east coast of Northern Honshu. It bore the brunt of the disaster, with ships, cars, buildings and people all smashed together and then sucked out to sea. In the days that followed, it was the Nuclear Power Plant in Fukushima that was the focus of attention. A huge radiation zone was created around the plant as the core melted down. This released radiation into the air, water and the land nearby. Unfortunately, the impact of the meltdown appears to have initially been covered up by the operating company and to some extent, the government.

More recently, however, publicity arose as the government was prompted to take over the clean-up task. As of this writing, it is still a huge

---

7   World Health Organization Press Release, "IARC Classifies Radiofrequency Electromagnetic Fields as Possibly Carcinogenic to Humans." Released May, 2011.

ongoing effort with very limited success. The government attempted to permanently freeze a large amount of the ground around the plant to prevent highly radioactive water from entering the ocean. Meanwhile the air continues to be polluted with radioactive cesium. The jet stream is such that the US, although a long way from Japan, is downwind and will bear the effect of the contamination for years to come.

It was announced recently, that deadly Fukushima beta radiation levels are now up thousands of times higher across all of North America than they were just a year ago. Beta radiation is actually a stream of high speed electrons. These high energy electrons carry a negative charge, and are sent out by some radioactive nuclei when they decompose during nuclear reactions. It should be noted that radiation of any sort can't be removed by distillation, but the radioactive elements emitting the radiation *can* be removed by distillation.

Dr. Russell Blaylock, a neurosurgeon and natural health expert, recently stated in a popular health magazine, *Newsmax Health*, that many Americans are getting more radiation today from three major sources "The three major sources are the Japanese nuclear reactor leak, nuclear power plants, and radon emission from earth." "Where you live will determine a lot of your yearly exposure," according to Dr. Blaylock.

"If you're living on the West Coast," he went on to say, "you're already getting more exposure from the Japanese reactor accident than is good for your health, as well as if you live within a mile of a nuclear power plant, or have a home built near high radon-emitting rocks … There is a lot of variability within the United States." he said.[8]

By now, most of us have been warned about the potential dangers of breathing and drinking radon gas. Radon is the radioactive by-product from the decay of radium and is a natural phenomenon in many parts

---

8   Russell Blaylock, "Japanese Radiation Could Pose Threat for US," Newsmax Health, March 15, 2011.

of the country. It is particularly prevalent in the Midwestern parts of the United States where it can be a point-of-entry problem for both air and water.

Chapter 7:

Taking Control of Your Water Supply

"Water, water everywhere, nor any drop to drink."

Samuel Taylor Coleridge

Efforts to combat water pollution have been made for as long as the problem itself has existed. Collecting rain water and melting snow were once popular ways to secure pure, soft water for drinking and household purposes. Today, however, we know the increased number of atmospheric pollutants negate most of the cleansing effects of the natural hydrologic cycle.

Boiling water to kill germs is also considered a purification technique by most people. Boiling, of course, does kill germs, but does not remove the microscopic remains from the water. Boiling will also drive off other organic chemicals that vaporize at a temperature lower than the boiling point of water. It will not, however, remove organic substances with boiling points higher than that of water, nor any inorganic minerals. In fact, merely boiling the water can increase the concentration of such substances.

As stated previously, early attempts to control waterborne diseases in drinking water used by an entire population involved the addition of chlorine, hypochlorite, iodine or ozone to the drinking water supply. These inorganic chemicals, in certain prescribed amounts, effectively killed bacteria without noticeably harming humans. In addition, filtration of water through a porous substance, such as sand, was found to remove certain other visible impurities. The quality of water varied from one area to another, with some naturally soft due to the absence of certain minerals, some naturally hard because of the presence of minerals, and some subject to specific, identifiable pollutants of a local industry. As a result, the treatment of drinking water supplies varied from one area to another and was generally the responsibility of a town or city government. As water treatment techniques evolved, local treatment facilities were constructed and certain standard procedures became common. The following are some examples of these procedures:

- Aerating the water to release trapped gases and to increase the oxygen content, thus improving the taste.

- Allowing the water to stand while undissolved particles of clay and silt settled to the bottom.

- Adding one or more jelly-like chemicals to attract undissolved particles that didn't settle in the previous step. As these particles gather and form clumps, they too settle to the bottom.

- Filtering the water through such substances as sand, coal or carbon to strain out any remaining particles.

- Removing the minerals that cause hardness by adding an appropriate chemical. Together they form an insoluble precipitate which is then removed from the water by filtration.

- Filtering out impurities that discolor water using finely powdered activated charcoal.

- Disinfecting the water, usually with chlorine, to kill bacteria.

Some water treatment plants disinfect the water as both a first and last step of treatment. If fluoride is to be added to the water supply, it is done after the final disinfectant process. Most of our country's municipal treatment facilities are well-equipped to perform these procedures, having been constructed soon after the processes became established in the 1920s and 1930s. Today, however, many of these treatment procedures and the facilities themselves are considered inadequate for satisfactorily treating the types of agricultural and industrial pollutants that now exist.

WHY ARE THERE CONTROVERSIES?

While the federal government has always insisted that the nation's local drinking water supplies be safe, pleasant-tasting, plentiful and low in cost, the mounting pollution problems of our modern era have forced it to play a greater role in regulating water quality. There are a number of agencies now responsible for monitoring the safety of our water supplies, including local, state and federal public health departments, the Food and Drug Administration and the Environmental Protection Agency. A number of legislative acts have also been passed to help deal with the problem. These include the Safe Drinking Water Act, the Federal Insecticide, Fungicide and Rodenticide Act, the Consumer Product Safety Act, the Pure Food and Cosmetic Act and the Toxic Substance Control Act.

While the specific regulations and laws set forth by these agencies are very complex indeed, they will invariably require cities to upgrade existing water treatment facilities to handle the growing influx of new pollutants. In many areas, local compliance with new regulations has been difficult or impossible due to the reluctance of local governments to levy additional taxes to pay for the upgrading and/or the regulations may not apply equally, or at all, to all areas of the country. Large population centers may be subject to one set of standards, while treatment plants serving smaller communities may be exempt, even though residents in those smaller communities may face the same, or worse, water pollution problems.

While the pre-treatment of public water before it is made accessible for drinking has been widely practiced in this country for more than a century, post-treatment of used water before it re-enters the water supply is relatively new. Methods of combating pollution at its source have been developed and used somewhat successfully in the treatment of public sewage and industrial waste water. However, in many cases, the money as well as the technological knowledge required to treat certain new pollutants is still lacking.

While all of us acknowledge the need for strict public regulation to maintain safe water supplies, some critics of the specific quality standards as set forth by the government note they are too lenient on the one hand and too stringent on the other. In other words, they claim water should be treated according to the way in which it is going to be used. Must the water used to fight fires undergo the same expensive cleansing procedures as the water we drink? And what of the water we drink? Are the results of the costly treatment at the local treatment site negated as the water passes through the worn and outdated water mains that pipe it to our homes? If treatment facilities are to be upgraded, shouldn't the entire water distribution networks in our cities and towns be upgraded too?

Of course these questions are expensive ones to answer. Ideally, there should be three grades of water available for different water uses: 1) utility grade for fire-fighting, watering lawns and flushing toilets; 2) work grade for household uses such as bathing and washing clothes; and, 3) drinking and cooking grade. For everyone to have these three grades of water would require untold amounts of money if each were to be piped separately to the general public from the water company. Also, the upgrading of water mains wouldn't ensure the purity of our tap water unless we replaced outdated plumbing systems in our individual homes as well. New and different approaches to solving the water pollution problem are presented almost daily, but the cost and practicality of many preclude their implementation on a large scale.

IT'S TIME FOR INDIVIDUAL ACTION!

The government isn't the only one attempting to develop workable solutions to the water problems around the world. During the last several decades, the private sector has made some significant inroads, most notably in the area of improving the quality of the drinking water that comes out of our taps. Most of them have one or more benefits and deserve your thoughtful consideration. This is where home water distillers have come in.

Twenty years ago there were a number of American manufacturers of home water distillers, several of them water-cooled. Water-cooled distillers have many of the same drawbacks as reverse osmosis systems, another home water treatment that has grown in popularity.

Water-cooled distillers, or R.O.s, not only waste a good deal of water in the process of producing product water, but they also limit the posible installation locations. Not only that, they make it much harder to eliminate volatile gases from the water they treat. They rely exclusively on activated carbon filters to remove volatile organic compounds, or V.O.C.s. Water cooled home distillers are pretty much a thing of the past, but reverse osmosis systems are still quite popular, even though they are not nearly as user-friendly as distillation systems and are more costly to maintain. Air-cooled distillation is the way to go, as they waste very little water.

Because a water distiller boils the water to destroy bacteria and viruses, then collects the pure steam in a separate chamber, distillation is the optimal method to treat water at home in an emergency, as recommended by the Red Cross and FEMA. People who have done their homework will soon realize that distillation is considered by most to be the gold standard of water purification. It has the highest percentage of removal on the broadest spectrum of contaminants and it does it consistently. It is also done at a small fraction of the cost of buying bottled water. This is why many doctors prescribe distilled water for their patients.

Unfortunately, a growing and irritating number of unscrupulous entrepreneurs and businesses are attracted to the water purification industry. These Johnny-Come-Lately enterprises are seeking to entice unwary consumers to products that fail to deliver the exaggerated health benefits that they claim and promise. Now for some good news. As stated previously, there is an extensive and well-documented website that is very informative and effective in rebutting the merits of these miracle solutions to drinking water problems. It is written by a retired chemistry professor and I would encourage you to check it out. It can be found at: http://www.chem1.com/CQ/gallery.html. Or if you type the following words into a search engine, "gallery of water related pseudoscience", you will also find the list. It will uncover the truth about water from a purely scientific perspective that is easy to understand. I believe you will be amazed at the wealth of useful information that is available there to untangle the water-related scams and protect today's confused and unsuspecting consumers.

BOTTLED WATER IS ONE OPTION

The first option is to buy bottled water and hundreds of thousands of people are now doing exactly that, especially in those areas where pollution problems have been widely publicized. Bottled water, however, can vary almost radically in quality. To date there are no uniform regulations governing the quality of domestic bottled water.

Even those labeled as natural mountain or natural spring water have been subjected to a variety of natural pollutants and must be treated in some way prior to bottling. Besides, buying bottled water can be very expensive and is usually pretty inconvenient. Freshness can also be a questionable factor being affected by both the length of time it has been stored in the plastic bottle and the chemical composition of the plastic bottle itself. Not to mention the environmental consequences of using so many plastic bottles. It has been found that eight out of 10 plastic bottles end up in the landfill and each take over 1,000 years to biodegrade.

OTHER ALTERNATIVES ARE AVAILABLE

Other alternatives to drinking ordinary tap water have come on the market so quickly that most of us are unfamiliar with the processes themselves and not really sure of the expected results. Most of the popular home water purification methods involve some type of filtering device.

"This filter is not intended to purify water"

A disclaimer written on a water filter box from a reputable brand name in the filtration industry.

Filters can remove select contaminants from water, using materials such as a screen, a net, paper or most commonly, activated charcoal. Activated charcoal will attract certain select particles from the water until it is saturated. Charcoal filters will adsorb some dissolved gases, light organic compounds and can drastically improve the color of water. They do not, however, affect most dissolved solids such as metals. Nor do they necessarily remove bacteria or viruses. If the filter is allowed to become saturated with a concentration of impurities, it can actually become a breeding ground for more bacteria, even when impregnated with silver or silver compounds.

Reverse osmosis filters attempt to improve water by separating it from dissolved minerals and chemicals using a semi-permeable membrane, usually manufactured from an organic substance similar to cling wrap or cellophane. Some organic solvents will react with the membrane or pass right through it.

Reverse osmosis (RO) membranes do substantially reduce all suspended and dissolved matter in water, but are affected by a wide range of variables including temperature, pressure and pH, just to name a few. They also waste a substantial amount of water to a drain. The process is generally not as effective in demineralizing water as is distillation, nor does it necessarily guarantee removal of bacteria. Bacteria can in fact grow through an RO membrane, which is then befouled. Iron and other hard minerals can also collect on RO membranes, which then require descaling.

DISTILLATION IS THE BEST ALTERNATIVE

In the opinion of many, the single most effective method of home water purification is distillation combined with post-filtration through an activated charcoal filter. Modern distillers are stainless steel appliances with many convenience features.

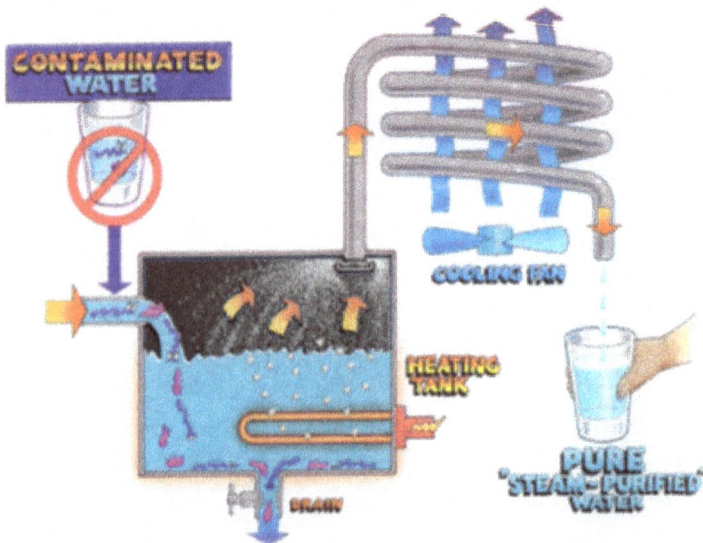

Distillers simulate the natural hydrologic cycle in three steps, first by boiling the water. This high prolonged heat kills and disintegrates all types of micro-organisms: bacteria, protozoa and viruses. Second, as the water is vaporized, the disintegrated microorganisms, as well as most chemicals and minerals, are left behind in the boiling chamber. Third, the steam is then condensed in a sterile condensation coil or chamber and reforms as high-purity, freshly distilled water.

Distillation is a full spectrum solution when used in combination with an activated carbon filter. Distillers typically require about three kilowatt hours (KWH) of electricity to produce one gallon of distilled water. The boiling chamber requires regular descaling if used on hard water. Softened water is definitely recommended for a distiller. The sodium which is added to water during the artificial softening process will be removed almost completely.

DISTILLATION HAS MANY SIDE BENEFITS

Merely filtering the water is like using a Band-Aid. Distillation, on the other hand, is more like doing surgery on the water. It not only improves the taste and odor of the tap water, but assures the ultimate safety of personal drinking and cooking water. Furthermore, distilled water is found to be a more acceptable drinking water for dogs, cats and other pets. Many have discovered that using distilled water for watering houseplants results in more luscious growth and longer-lasting blooms.

Distillers are available in a wide range of sizes for residential and commercial uses. They are the next major appliance for inclusion in the modern home. Besides the added convenience, freshness and quality people enjoy from owning a state-of-the-art home distillation system, they have eliminated the expense of regularly purchasing bottled water. No longer is distilled water seen as only a necessity for steam irons and contact lenses, but for a full range of personal and household uses. Water of enhanced quality has suddenly become a necessity, not a luxury.

EXPERT OPINIONS ON DISTILLED WATER.

"The only type of water that seems to be fit for consumption is distilled water…Distillation is the single most effective method of water purification."

— **Peter A. Lodewick, M.D.,** *A Diabetic Doctor Looks at Diabetes*

"We believe that only steam-distilled water should be consumed."

— **James F. Balch, MD. & Phyllis A. Balch, C.N.C.,** *Prescription for Nutritional Healing*

"Distilled water is the world's best and purest water!"

— **Dr. Paul C. Bragg & Dr. Patricia Bragg,** *Water-The Shocking Truth*

"If you decide on bottled water, make sure it's distilled, (however), in the long run you'll save money if you clean your water at home."

— **Dr. Robert D. Willix, Jr., M.D.,** *Maximum Health*

"Distilled water is the only water which is pure - the only water free from all impurities."

— **Dr. Allen E. Banik,** *The Choice Is Clear*

"I use a steam distiller to purify my water, because I believe distilled is the cleanest you can get."

— **Dr. Andrew Weil, The Healthy Kitchen**

"Even tap water invariably contains a variety of poisons such as chlorine, chloramine, asbestos, pesticides, fluoride, copper, mercury, and lead. The best way to remove all these contaminants is by distilling."

— **Dr. David Kennedy, D.D.S.,** *How to Save Your Teeth: Toxic-Free Preventive Dentistry*

"The home distiller is the best method and also the best way to get distilled water. It is the only reliable home water purification for taking fluoride out of the water."

— Dr. John Yiamoyuiannis, Ph.D., *Fluoride: The Aging Factor*

"Far and away the cleanest water is produced by the new home distillers."

— Dr. Michael Colgan, Ph.D., CCN, Optimum Sports Nutrition

"Water hardness [inorganic minerals in solution] is the underlying cause of many, if not all, of the diseases resulting from poisons in the intestinal tract. These [hard minerals] pass from the intestinal walls and get into the lymphatic system, which delivers all of its products to the blood, which in turn, distributes to all parts of the body. This is the cause of much human disease."

— Dr. Charles Mayo, Mayo Clinic

"The statement that distilled water leaches minerals from the body has no basis in fact. It doesn't leach out minerals that have become part of the cell structure. It can't and won't. It collects only minerals that have already been rejected or excreted by the cells…"

— Harvey & Marilyn Diamond, Fit for Life II: Living Health

WHERE DO YOU GO FROM HERE?

Judging by the popularity today of small bottled water brands such as Aquafina, Dasani and others, there is a large and ever-growing concern for the availability, convenience and quality of drinking water in this country. It is my understanding that water is now outselling soft drinks sold by Pepsi and Coca Cola. Who would have imagined 25 years ago that consumers would be willing to pay two dollars or more for a half-liter bottle of purified drinking water. (This works out to be around eight dollars a gallon!)

It is much more economical to make your own distilled water in the convenience of our own home. Using three KWH per gallon for energy consumption, this amounts to about 27 cents for a full gallon of distilled water in most places in the United States. (It is slightly more on both coasts.)

Obviously you are among the concerned if you are this far into the book. By reading this book, you have actually done a pretty thorough investigation of the fascinating world of water. By now I sincerely hope that you have managed to convince yourself about the wonderful health-giving benefits of owning a home water distillation system. But there's more...

IF YOU CARE—SHARE!

We would like to urge you to pass the health-giving information you have learned from this book on to others, especially to family members, close friends and any associates that you feel would benefit.

And so... it's entirely up to you. With your own Pure Water™ distiller, you can experience better living.

ABOUT THE AUTHOR

Eldon C. Muehling, a native Nebraskan, is a graduate of the University of Nebraska at Kearney. While there, he taught college-level mathematics. He was a high school chemistry and physics teacher for 14 years. During the summers he attended eight different Colleges and Universities where he studied various fields of physical and biological sciences using stipends from the National Science Foundation. He acquired more than 70 graduate hours in chemistry, physics, earth science and radiation biology. He earned his Master's Degree in Science Education from Northwest Missouri State University in 1970.

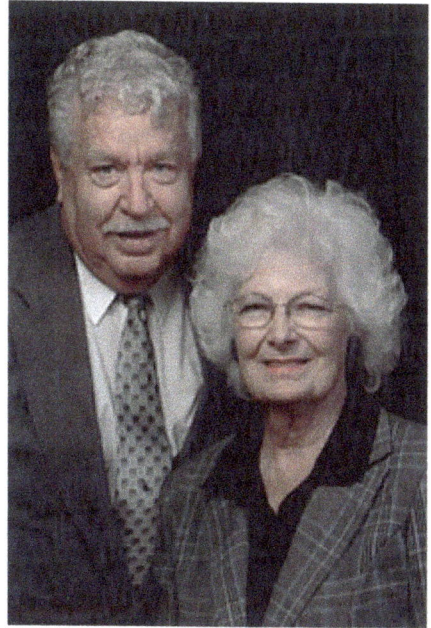

Eldon's life-long interest in water dates back to his first college course in chemistry. He has always enjoyed sharing his knowledge of science with others who showed interest. Those who know him and have been helped by him, often refer to him affectionately as "Dr. Water."

Eldon has worked in the water industry since 1977. He has been a guest on a number radio and TV Shows in several states, and written articles for several trade magazines, and edited and authored a number of books about water including *You Have a Right to Know*, *Water for the Eighties… A Cause For Concern,* and *Pure Water Now: A Time For Action*. This book, *Pure Water for Better Living,* is an expanded and revised version of that publication.

Eldon has been married to Nancy for 57 plus years. They enjoy their three children, seven grandchildren and one great granddaughter. Eldon and Nancy both enjoy landscaping and gardening.

www.ingramcontent.com/pod-product-compliance
Lightning Source LLC
Chambersburg PA
CBHW041300040426
42334CB00028BA/3103